An Elementary Course in Synthetic Projective Geometry

Derrick Norman Lehmer

Classical Prints

This edition published in 2022

ISBN : 9789356536562

Contents

Preface

The following course is intended to give, in as simple a way as possible, the essentials of synthetic projective geometry. While, in the main, the theory is developed along the well-beaten track laid out by the great masters of the subject, it is believed that there has been a slight smoothing of the road in some places. Especially will this be observed in the chapter on Involution. The author has never felt satisfied with the usual treatment of that subject by means of circles and anharmonic ratios. A purely projective notion ought not to be based on metrical foundations. Metrical developments should be made there, as elsewhere in the theory, by the introduction of infinitely distant elements.

The author has departed from the century-old custom of writing in parallel columns each theorem and its dual. He has not found that it conduces to sharpness of vision to try to focus his eyes on two things at once. Those who prefer the usual method of procedure can, of course, develop the two sets of theorems side by side; the author has not found this the better plan in actual teaching.

As regards nomenclature, the author has followed the lead of the earlier writers in English, and has called the system of lines in a plane which all pass through a point a *pencil of rays* instead of a *bundle of rays*, as later writers seem inclined to do. For a point considered as made up of all the lines and planes through it he has ventured to use the term *point system*, as being the natural dualization of the usual term *plane system.* He has also rejected the term *foci of an involution*, and has not used the customary terms for classifying involutions—*hyperbolic involution*, *elliptic involution* and *parabolic involution.* He has found that all these terms are very confusing to the student, who inevitably tries to connect them in some way with the conic sections.

Enough examples have been provided to give the student a clear grasp of the theory. Many are of sufficient generality to serve as a basis for individual investigation on the part of the student. Thus, the third example at the end of the first chapter will be found to be very fruitful in interesting results. A correspondence is there indicated between lines in space and circles through a fixed point in space. If the student will trace a few of the consequences of that correspondence, and determine what configurations of circles correspond to intersecting lines, to lines in a plane, to lines of a plane pencil, to lines cutting three skew lines, etc., he will have acquired no little practice in picturing to himself figures in space.

The writer has not followed the usual practice of inserting historical notes at the foot of the page, and has tried instead, in the last chapter, to give a consecutive account of the history of pure geometry, or, at least, of as much of it as the student will be able to appreciate who has mastered the course as given in the preceding chapters. One is not apt to get a very wide view of the history of a subject by reading a hundred biographical footnotes, arranged in no sort of sequence. The writer, moreover, feels that the proper time to learn the history of a subject is after the student has some general ideas of the subject itself.

The course is not intended to furnish an illustration of how a subject may be developed, from the smallest possible number of fundamental assumptions. The author is aware of the importance of work of this sort, but he does not believe it is possible at the present time to write a book along such lines which shall be of much use for elementary students. For the purposes of this course the student should have a thorough grounding in ordinary elementary geometry so far as to include the study of the circle and of similar triangles. No solid geometry is needed beyond the little used in the proof of Desargues' theorem (25), and, except in certain metrical developments of the general theory, there will be no call for a knowledge of trigonometry or analytical geometry. Naturally the student who is equipped with these subjects as well as with the calculus will be a little more mature, and may be expected to follow the course all the more easily. The author has had no difficulty, however, in presenting it to students in the freshman class at the University of California.

The subject of synthetic projective geometry is, in the opinion of the writer, destined shortly to force its way down into the secondary schools; and if this little book helps to accelerate the movement, he will feel amply repaid for the task of working the materials into a form available for such schools as well as for the lower classes in the university.

The material for the course has been drawn from many sources. The author is chiefly indebted to the classical works of Reye, Cremona, Steiner, Poncelet, and Von Staudt. Acknowledgments and thanks are also due to Professor Walter C. Eells, of the U.S. Naval Academy at Annapolis, for his searching examination and keen criticism of the manuscript; also to Professor Herbert Ellsworth Slaught, of The University of Chicago, for his many valuable suggestions, and to Professor B. M. Woods and Dr. H. N. Wright, of the University of California, who have tried out the methods of presentation, in their own classes.

D. N. LEHMER

BERKELEY, CALIFORNIA

CHAPTER I
ONE-TO-ONE CORRESPONDENCE

1. Definition of one-to-one correspondence. Given any two sets of individuals, if it is possible to set up such a correspondence between the two sets that to any individual in one set corresponds one and only one individual in the other, then the two sets are said to be in *one-to-one correspondence* with each other. This notion, simple as it is, is of fundamental importance in all branches of science. The process of counting is nothing but a setting up of a one-to-one correspondence between the objects to be counted and certain words, 'one,' 'two,' 'three,' etc., in the mind. Many savage peoples have discovered no better method of counting than by setting up a one-to-one correspondence between the objects to be counted and their fingers. The scientist who busies himself with naming and classifying the objects of nature is only setting up a one-to-one correspondence between the objects and certain words which serve, not as a means of counting the objects, but of listing them in a convenient way. Thus he may be able to marshal and array his material in such a way as to bring to light relations that may exist between the objects themselves. Indeed, the whole notion of language springs from this idea of one-to-one correspondence.

2. Consequences of one-to-one correspondence. The most useful and interesting problem that may arise in connection with any one-to-one correspondence is to determine just what relations existing between the individuals of one assemblage may be carried over to another assemblage in one-to-one correspondence with it. It is a favorite error to assume that whatever holds for one set must also hold for the other. Magicians are apt to assign magic properties to many of the words and symbols which they are in the habit of using, and scientists are constantly confusing objective things with the subjective formulas for them. After the physicist has set up correspondences between physical facts and mathematical formulas, the "interpretation" of these formulas is his most important and difficult task.

3. In mathematics, effort is constantly being made to set up one-to-one correspondences between simple notions and more complicated ones, or between the well-explored fields of research and fields less known. Thus, by means of the mechanism employed in analytic geometry, algebraic theorems are made to yield geometric ones, and vice versa. In geometry we get at the properties of the conic sections by means of the properties of the straight line, and cubic surfaces are studied by means of the plane.

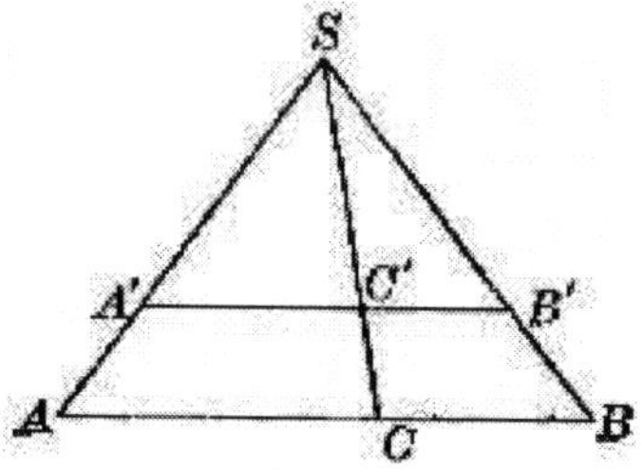

FIG. 1

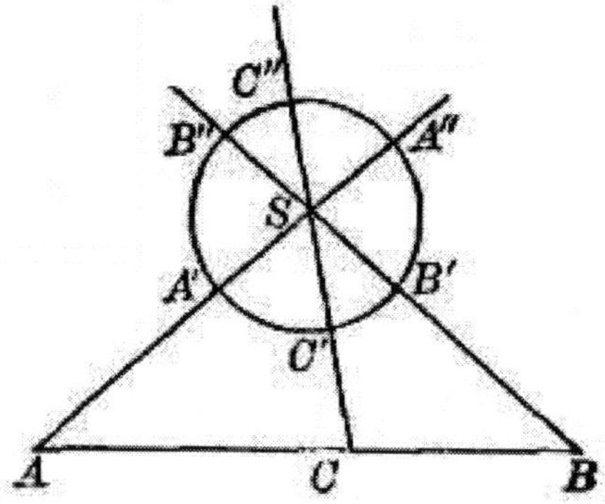

FIG. 2

4. One-to-one correspondence and enumeration. If a one-to-one correspondence has been set up between the objects of one set and the objects of another set, then the inference may usually be drawn that they have the same number of elements. If, however, there is an infinite number of individuals in each of the two sets, the notion of counting is necessarily ruled out. It may be possible, nevertheless, to set up a one-to-one correspondence between the elements of two sets even when the number is infinite. Thus, it is easy to set up such a correspondence between the points of a line an inch long and the points of a line two inches long. For let the lines (Fig. 1) be AB and $A'B'$. Join AA' and BB', and let these joining lines meet in S. For every point C on AB a point C' may be found on $A'B'$ by joining C to S and noting the point C' where CS meets $A'B'$. Similarly, a point C may be found on AB for any point C' on $A'B'$. The correspondence is clearly one-to-one, but it would be absurd to infer from this that there were just as many points on AB as on $A'B'$. In fact, it would be just as reasonable to infer that there were twice as many points on $A'B'$ as on AB. For if we bend $A'B'$ into a circle with center at S (Fig. 2), we see that for every point C on AB there are two points on $A'B'$. Thus it is seen that the notion of one-to-one correspondence is more extensive than the notion of counting, and includes the notion of counting only when applied to finite assemblages.

5. Correspondence between a part and the whole of an infinite assemblage. In the discussion of the last paragraph the remarkable fact was brought to light that it is sometimes possible to set the elements of an assemblage into one-to-one correspondence with a part of those elements. A moment's reflection will convince one that this is never possible when there is a finite number of elements in the assemblage.—Indeed, we may take this property as our definition of an infinite assemblage, and say that an infinite assemblage is one that may be put into one-to-one correspondence with part of itself. This has the advantage of being a positive definition, as opposed to the usual negative definition of an infinite assemblage as one that cannot be counted.

6. Infinitely distant point. We have illustrated above a simple method of setting the points of two lines into one-to-one correspondence. The same illustration will serve also to show how it is possible to set the points on a line into one-to-one correspondence with the lines through a point. Thus, for any point C on the line AB there is a line SC through S. We must assume the line AB extended indefinitely in both directions, however, if we are to have a point on it for every line through S; and even with this extension there is one line through S, according to Euclid's postulate, which does not meet the line AB and which therefore has no point on AB to correspond to it. In order to smooth out this discrepancy we are accustomed to assume the existence of an *infinitely distant* point on the line AB and to assign this point as the corresponding point of the exceptional line of S. With this understanding, then, we may say that we have set the lines through a point and the points on a line into one-to-one correspondence. This correspondence is of such fundamental importance in the study of projective geometry that a special name is given to it. Calling the totality of points on a line a *point-row*, and the totality of lines through a point a *pencil of rays*, we say that the point-row and the pencil related as above are in *perspective position*, or that they are *perspectively related.*

7. Axial pencil; fundamental forms. A similar correspondence may be set up between the points on a line and the planes through another line which does not meet the first. Such a system of planes is called an *axial pencil*, and the three assemblages—the point-row, the pencil of rays, and the axial pencil—are called *fundamental forms.* The fact that they may all be set into one-to-one correspondence with each other is expressed by saying that they are of the same order. It is usual also to speak of them as of the first order. We shall see presently that there are other assemblages which cannot be put into this sort of one-to-one correspondence with the points on a line, and that they will very reasonably be said to be of a higher order.

8. Perspective position. We have said that a point-row and a pencil of rays are in perspective position if each ray of the pencil goes through the point of the point-row which corresponds to it. Two pencils of rays are also said to be in perspective position if corresponding rays meet on a straight line which is called the axis of perspectivity. Also, two point-rows are said to be in perspective position if corresponding points lie on straight lines through a point which is called the center of perspectivity. A point-row and an axial pencil are in perspective position if each plane of the pencil goes through the point on the point-row which corresponds to it, and an axial pencil and a pencil of rays are in perspective position if each ray lies in the plane which corresponds to it; and, finally, two axial pencils are perspectively related if corresponding planes meet in a plane.

9. Projective relation. It is easy to imagine a more general correspondence between the points of two point-rows than the one just described. If we take two perspective pencils, *A* and *S*, then a point-row *a* perspective to *A* will be in one-to-one correspondence with a point-row *b* perspective to *B*, but corresponding points will not, in general, lie on lines which all pass through a point. Two such point-rows are said to be *projectively related*, or simply projective to each other. Similarly, two pencils of rays, or of planes, are projectively related to each other if they are perspective to two perspective point-rows. This idea will be generalized later on. It is important to note that between the elements of two projective fundamental forms there is a one-to-one correspondence, and also that this correspondence is in general *continuous*; that is, by taking two elements of one form sufficiently close to each other, the two corresponding elements in the other form may be made to approach each other arbitrarily close. In the case of point-rows this continuity is subject to exception in the neighborhood of the point "at infinity."

10. Infinity-to-one correspondence. It might be inferred that any infinite assemblage could be put into one-to-one correspondence with any other. Such is not the case, however, if the correspondence is to be continuous, between the points on a line and the points on a plane. Consider two lines which lie in different planes, and take *m* points on one and *n* points on the other. The number of lines joining the *m* points of one to the *n* points jof the other is clearly *mn*. If we symbolize the totality of points on a line by [infinity], then a reasonable symbol for the totality of lines drawn to cut two lines would be [infinity]^2. Clearly, for every point on one line there are [infinity] lines cutting across the other, so that the correspondence might be called [infinity]-to-one. Thus the assemblage of lines cutting across two lines is of higher order than the assemblage of points on a line; and as we have called the point-row an assemblage of the first order, the system of lines cutting across two lines ought to be called of the second order.

11. Infinitudes of different orders. Now it is easy to set up a one-to-one correspondence between the points in a plane and the system of lines cutting across two lines which lie in different planes. In fact, each line of the system of lines meets the plane in one point, and each point in the plane determines one and only one line cutting across the two given lines—namely, the line of intersection of the two planes determined by the given point with each of the given lines. The assemblage of points in the plane is thus of the same order as that of the lines cutting across two lines which lie in different planes, and ought therefore to be spoken of as of the second order. We express all these results as follows:

12. If the infinitude of points on a line is taken as the infinitude of the first order, then the infinitude of lines in a pencil of rays and the infinitude of planes in an axial pencil are also of the first order, while the infinitude of lines cutting across two "skew" lines, as well as the infinitude of points in a plane, are of the second order.

13. If we join each of the points of a plane to a point not in that plane, we set up a one-to-one correspondence between the points in a plane and the lines through a point in space. *Thus the infinitude of lines through a point in space is of the second order.*

14. If to each line through a point in space we make correspond that plane at right angles to it and passing through the same point, we see that *the infinitude of planes through a point in space is of the second order.*

15. If to each plane through a point in space we make correspond the line in which it intersects a given plane, we see that *the infinitude of lines in a plane is of the second order.* This may also be seen by setting up a one-to-one correspondence between the points on a plane and the lines of that plane. Thus, take a point S not in the plane. Join any point M of the plane to S. Through S draw a plane at right angles to MS. This meets the given plane in a line m which may be taken as corresponding to the point M. Another very important method of setting up a one-to-one correspondence between lines and points in a plane will be given later, and many weighty consequences will be derived from it.

16. Plane system and point system. The plane, considered as made up of the points and lines in it, is called a *plane system* and is a fundamental form of the second order. The point, considered as made up of all the lines and planes passing through it, is called a *point system* and is also a fundamental form of the second order.

17. If now we take three lines in space all lying in different planes, and select *l* points on the first, *m* points on the second, and *n* points on the third, then the total number of planes passing through one of the selected points on each line will be *lmn*. It is reasonable, therefore, to symbolize the totality of planes that are determined by the [infinity] points on each of the three lines by [infinity]^3, and to call it an infinitude of the *third* order. But it is easily seen that every plane in space is included in this totality, so that *the totality of planes in space is an infinitude of the third order.*

18. Consider now the planes perpendicular to these three lines. Every set of three planes so drawn will determine a point in space, and, conversely, through every point in space may be drawn one and only one set of three planes at right angles to the three given lines. It follows, therefore, that *the totality of points in space is an infinitude of the third order.*

19. Space system. Space of three dimensions, considered as made up of all its planes and points, is then a fundamental form of the *third* order, which we shall call a *space system.*

20. Lines in space. If we join the twofold infinity of points in one plane with the twofold infinity of points in another plane, we get a totality of lines of space which is of the fourth order of infinity. *The totality of lines in space gives, then, a fundamental form of the fourth order.*

21. Correspondence between points and numbers. In the theory of analytic geometry a one-to-one correspondence is assumed to exist between points on a line and numbers. In order to justify this assumption a very extended definition of number must be made use of. A one-to-one correspondence is then set up between points in the plane and pairs of numbers, and also between points in space and sets of three numbers. A single constant will serve to define the position of a point on a line; two, a point in the plane; three, a point in space; etc. In the same theory a one-to-one correspondence is set up between loci in the plane and equations in two variables; between surfaces in space and equations in three variables; etc. The equation of a line in a plane involves two constants, either of which may take an infinite number of values. From this it follows that there is an infinity of lines in the plane which is of the second order if the infinity of points on a line is assumed to be of the first. In the same way a circle is determined by three conditions; a sphere by four; etc. We might then expect to be able to set up a one-to-one correspondence between circles in a plane and points, or planes in space, or between spheres and lines in space. Such, indeed, is the case, and it is often possible to infer theorems concerning spheres from theorems concerning lines, and vice versa. It is possibilities such as these that, give to the theory of one-to-one correspondence its great importance for the mathematician. It must not be forgotten, however, that we are considering only *continuous* correspondences. It is perfectly possible to set, up a one-to-one correspondence between the points of a line and the points of a plane, or, indeed, between the points of a line and the points of a space of any finite number of dimensions, if the correspondence is not restricted to be continuous.

22. Elements at infinity. A final word is necessary in order to explain a phrase which is in constant use in the study of projective geometry. We have spoken of the "point at infinity" on a straight line—a fictitious point only used to bridge over the exceptional case when we are setting up a one-to-one correspondence between the points of a line and the lines through a point. We speak of it as "a point" and not as "points," because in the geometry studied by Euclid we assume only one line through a point parallel to a given line. In the same sense we speak of all the points at infinity in a plane as lying on a line, "the line at infinity," because the straight line is the simplest locus we can imagine which has only one point in common with any line in the plane. Likewise we speak of the "plane at infinity," because that seems the most convenient way of imagining the points at infinity in space. It must not be inferred that these conceptions have any essential connection with physical facts, or that other means of picturing to ourselves the infinitely distant configurations are not possible. In other branches of mathematics, notably in the theory of functions of a complex variable, quite different assumptions are made and quite different conceptions of the elements at infinity are used. As we can know nothing experimentally about such things, we are at liberty to make any assumptions we please, so long as they are consistent and serve some useful purpose.

PROBLEMS

1. Since there is a threefold infinity of points in space, there must be a sixfold infinity of pairs of points in space. Each pair of points determines a line. Why, then, is there not a sixfold infinity of lines in space?

2. If there is a fourfold infinity of lines in space, why is it that there is not a fourfold infinity of planes through a point, seeing that each line in space determines a plane through that point?

3. Show that there is a fourfold infinity of circles in space that pass through a fixed point. (Set up a one-to-one correspondence between the axes of the circles and lines in space.)

4. Find the order of infinity of all the lines of space that cut across a given line; across two given lines; across three given lines; across four given lines.

5. Find the order of infinity of all the spheres in space that pass through a given point; through two given points; through three given points; through four given points.

6. Find the order of infinity of all the circles on a sphere; of all the circles on a sphere that pass through a fixed point; through two fixed points; through three fixed points; of all the circles in space; of all the circles that cut across a given line.

7. Find the order of infinity of all lines tangent to a sphere; of all planes tangent to a sphere; of lines and planes tangent to a sphere and passing through a fixed point.

8. Set up a one-to-one correspondence between the series of numbers *1*, *2*, *3*, *4*, ... and the series of even numbers *2*, *4*, *6*, *8* .... Are we justified in saying that there are just as many even numbers as there are numbers altogether?

9. Is the axiom "The whole is greater than one of its parts" applicable to infinite assemblages?

10. Make out a classified list of all the infinitudes of the first, second, third, and fourth orders mentioned in this chapter.

CHAPTER II
RELATIONS BETWEEN FUNDAMENTAL FORMS IN ONE-TO-ONE CORRESPONDENCE WITH EACH OTHER

23. Seven fundamental forms. In the preceding chapter we have called attention to seven fundamental forms: the point-row, the pencil of rays, the axial pencil, the plane system, the point system, the space system, and the system of lines in space. These fundamental forms are the material which we intend to use in building up a general theory which will be found to include ordinary geometry as a special case. We shall be concerned, not with measurement of angles and areas or line segments as in the study of Euclid, but in combining and comparing these fundamental forms and in "generating" new forms by means of them. In problems of construction we shall make no use of measurement, either of angles or of segments, and except in certain special applications of the general theory we shall not find it necessary to require more of ourselves than the ability to draw the line joining two points, or to find the point of intersections of two lines, or the line of intersection of two planes, or, in general, the common elements of two fundamental forms.

24. Projective properties. Our chief interest in this chapter will be the discovery of relations between the elements of one form which hold between the corresponding elements of any other form in one-to-one correspondence with it. We have already called attention to the danger of assuming that whatever relations hold between the elements of one assemblage must also hold between the corresponding elements of any assemblage in one-to-one correspondence with it. This false assumption is the basis of the so-called "proof by analogy" so much in vogue among speculative theorists. When it appears that certain relations existing between the points of a given point-row do not necessitate the same relations between the corresponding elements of another in one-to-one correspondence with it, we should view with suspicion any application of the "proof by analogy" in realms of thought where accurate judgments are not so easily made. For example, if in a given point-row u three points, A, B, and C, are taken such that B is the middle point of the segment AC, it does not follow that the three points A', B', C' in a point-row perspective to u will be so related. Relations between the elements of any form which do go over unaltered to the corresponding elements of a form projectively related to it are called *projective relations*. Relations involving measurement of lines or of angles are not projective.

25. Desargues's theorem. We consider first the following beautiful theorem, due to Desargues and called by his name.

If two triangles, A, B, C and A', B', C', are so situated that the lines AA', BB', and CC' all meet in a point, then the pairs of sides AB and $A'B'$, BC and $B'C'$, CA and $C'A'$ all meet on a straight line, and conversely.

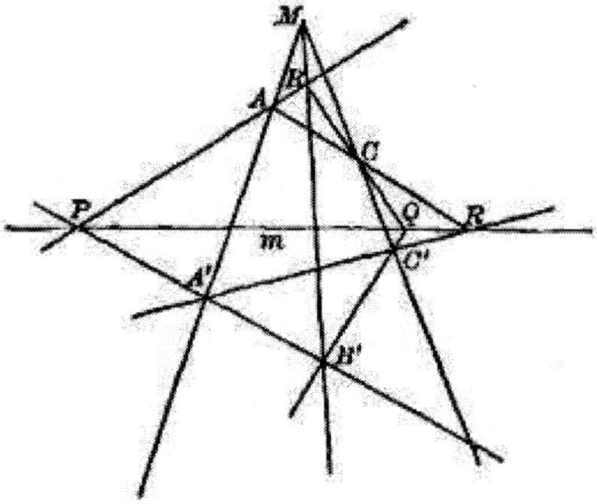

FIG. 3

Let the lines AA', BB', and CC' meet in the point M (Fig. 3). Conceive of the figure as in space, so that M is the vertex of a trihedral angle of which the given triangles are plane sections. The lines AB and $A'B'$ are in the same plane and must meet when produced, their point of intersection being clearly a point in the plane of each triangle and therefore in the line of intersection of these two planes. Call this point P. By similar reasoning the point Q of intersection of the lines BC and $B'C'$ must lie on this same line as well as the point R of intersection of CA and $C'A'$. Therefore the points P, Q, and R all lie on the same line *m*. If now we consider the figure a plane figure, the points *P*, *Q*, and *R* still all lie on a straight line, which proves the theorem. The converse is established in the same manner.

26. Fundamental theorem concerning two complete quadrangles. This theorem throws into our hands the following fundamental theorem concerning two complete quadrangles, a *complete quadrangle* being defined as the figure obtained by joining any four given points by straight lines in the six possible ways.

Given two complete quadrangles, K, L, M, N and K', L', M', N', so related that KL, K'L', MN, M'N' all meet in a point A; LM, L'M', NK, N'K' all meet in a point Q; and LN, L'N' meet in a point B on the line AC; then the lines KM and K'M' also meet in a point D on the line AC.

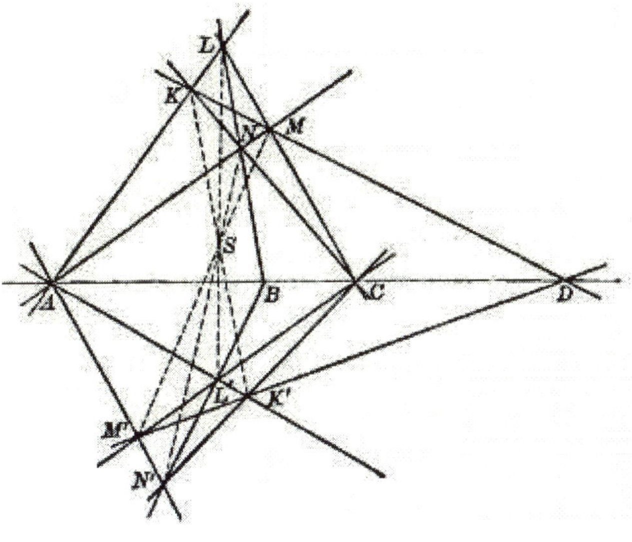

FIG. 4

For, by the converse of the last theorem, *KK'*, *LL'*, and *NN'* all meet in a point *S* (Fig. 4). Also *LL'*, *MM'*, and *NN'* meet in a point, and therefore in the same point *S*. Thus *KK'*, *LL'*, and *MM'* meet in a point, and so, by Desargues's theorem itself, *A*, *B*, and *D* are on a straight line.

27. Importance of the theorem. The importance of this theorem lies in the fact that, *A*, *B*, and *C* being given, an indefinite number of quadrangles *K′*, *L′*, *M′*, *N′* my be found such that *K′L′* and *M′N′* meet in *A*, *K′N′* and *L′M′* in *C*, with *L′N′* passing through *B*. Indeed, the lines *AK′* and *AM′* may be drawn arbitrarily through *A*, and any line through *B* may be used to determine *L′* and *N′*. By joining these two points to *C* the points *K′* and *M′* are determined. Then the line joining *K′* and *M′*, found in this way, must pass through the point *D* already determined by the quadrangle *K*, *L*, *M*, *N*. *The three points A, B, C, given in order, serve thus to determine a fourth point D.*

28. In a complete quadrangle the line joining any two points is called the *opposite side* to the line joining the other two points. The result of the preceding paragraph may then be stated as follows:

Given three points, *A*, *B*, *C*, in a straight line, if a pair of opposite sides of a complete quadrangle pass through *A*, and another pair through *C*, and one of the remaining two sides goes through *B*, then the other of the remaining two sides will go through a fixed point which does not depend on the quadrangle employed.

29. Four harmonic points. Four points, *A*, *B*, *C*, *D*, related as in the preceding theorem are called *four harmonic points*. The point *D* is called the *fourth harmonic of B with respect to A and C*. Since *B* and *D* play exactly the same rôle in the above construction, *B is also the fourth harmonic of D with respect to A and C*. *B* and *D* are called *harmonic conjugates with respect to A and C*. We proceed to show that *A* and *C* are also harmonic conjugates with respect to *B* and *D*—that is, that it is possible to find a quadrangle of which two opposite sides shall pass through *B*, two through *D*, and of the remaining pair, one through *A* and the other through *C*.

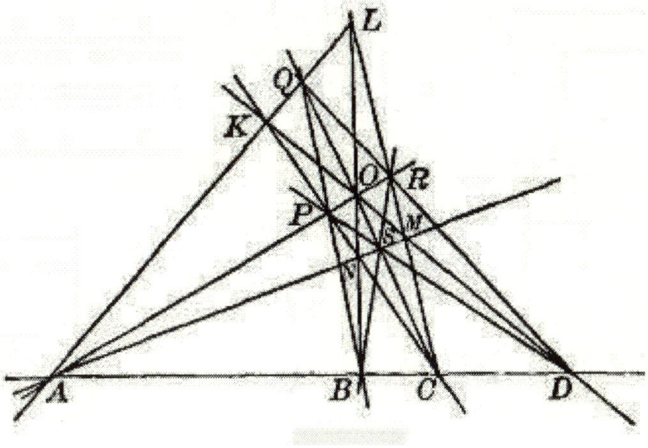

FIG. 5

Let O be the intersection of KM and LN (Fig. 5). Join O to A and C. The joining lines cut out on the sides of the quadrangle four points, P, Q, R, S. Consider the quadrangle P, K, Q, O. One pair of opposite sides passes through A, one through C, and one remaining side through D; therefore the other remaining side must pass through B. Similarly, RS passes through B and PS and QR pass through D. The quadrangle P, Q, R, S therefore has two opposite sides through B, two through D, and the remaining pair through A and C. A and C are thus harmonic conjugates with respect to B and D. We may sum up the discussion, therefore, as follows:

30. If A and C are harmonic conjugates with respect to B and D, then B and D are harmonic conjugates with respect to A and C.

31. Importance of the notion. The importance of the notion of four harmonic points lies in the fact that it is a relation which is carried over from four points in a point-row u to the four points that correspond to them in any point-row u' perspective to u.

To prove this statement we construct a quadrangle K, L, M, N such that KL and MN pass through A, KN and LM through C, LN through B, and KM through D. Take now any point S not in the plane of the quadrangle and construct the planes determined by S and all the seven lines of the figure. Cut across this set of planes by another plane not passing through S. This plane cuts out on the set of seven planes another quadrangle which determines four new harmonic points, A', B', C', D', on the lines joining S to A, B, C, D. But S may be taken as any point, since the original quadrangle may be taken in any plane through A, B, C, D; and, further, the points A', B', C', D' are the intersection of SA, SB, SC, SD by any line. We have, then, the remarkable theorem:

32. *If any point is joined to four harmonic points, and the four lines thus obtained are cut by any fifth, the four points of intersection are again harmonic.*

33. Four harmonic lines. We are now able to extend the notion of harmonic elements to pencils of rays, and indeed to axial pencils. For if we define *four harmonic rays* as four rays which pass through a point and which pass one through each of four harmonic points, we have the theorem

Four harmonic lines are cut by any transversal in four harmonic points.

34. Four harmonic planes. We also define *four harmonic planes* as four planes through a line which pass one through each of four harmonic points, and we may show that

Four harmonic planes are cut by any plane not passing through their common line in four harmonic lines, and also by any line in four harmonic points.

For let the planes α, β, γ, δ, which all pass through the line g, pass also through the four harmonic points A, B, C, D, so that α passes through A, etc. Then it is clear that any plane π through A, B, C, D will cut out four harmonic lines from the four planes, for they are lines through the intersection P of g with the plane π, and they pass through the given harmonic points A, B, C, D. Any other plane σ cuts g in a point S and cuts α, β, γ, δ in four lines that meet π in four points A', B', C', D' lying on PA, PB, PC, and PD respectively, and are thus four harmonic hues. Further, any ray cuts α, β, γ, δ in four harmonic points, since any plane through the ray gives four harmonic lines of intersection.

35. These results may be put together as follows:

Given any two assemblages of points, rays, or planes, perspectively related to each other, four harmonic elements of one must correspond to four elements of the other which are likewise harmonic.

If, now, two forms are perspectively related to a third, any four harmonic elements of one must correspond to four harmonic elements in the other. We take this as our definition of projective correspondence, and say:

36. Definition of projectivity. *Two fundamental forms are protectively related to each other when a one-to-one correspondence exists between the elements of the two and when four harmonic elements of one correspond to four harmonic elements of the other.*

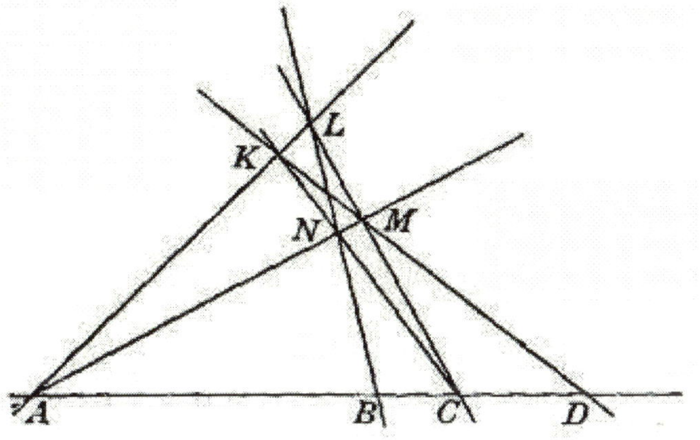

FIG. 6

37. Correspondence between harmonic conjugates. Given four harmonic points, *A*, *B*, *C*, *D*; if we fix *A* and *C*, then *B* and *D* vary together in a way that should be thoroughly understood. To get a clear conception of their relative motion we may fix the points *L* and *M* of the quadrangle *K*, *L*, *M*, *N* (Fig. 6). Then, as *B* describes the point-row *AC*, the point *N* describes the point-row *AM* perspective to it. Projecting *N* again from *C*, we get a point-row *K* on *AL* perspective to the point-row *N* and thus projective to the point-row *B*. Project the point-row *K* from *M* and we get a point-row *D* on *AC* again, which is projective to the point-row *B*. For every point *B* we have thus one and only one point *D*, and conversely. In other words, we have set up a one-to-one correspondence between the points of a single point-row, which is also a projective correspondence because four harmonic points *B* correspond to four harmonic points *D*. We may note also that the correspondence is here characterized by a feature which does not always appear in projective correspondences: namely, the same process that carries one from *B* to *D* will carry one back from *D* to *B* again. This special property will receive further study in the chapter on Involution.

38. It is seen that as *B* approaches *A*, *D* also approaches *A*. As *B* moves from *A* toward *C*, *D* moves from *A* in the opposite direction, passing through the point at infinity on the line *AC*, and returns on the other side to meet *B* at *C* again. In other words, as *B* traverses *AC*, *D* traverses the rest of the line from *A* to *C* through infinity. In all positions of *B*, except at *A* or *C*, *B* and *D* are separated from each other by *A* and *C*.

39. Harmonic conjugate of the point at infinity. It is natural to inquire what position of *B* corresponds to the infinitely distant position of *D*. We have proved (§ 27) that the particular quadrangle *K*, *L*, *M*, *N* employed is of no consequence. We shall therefore avail ourselves of one that lends itself most readily to the solution of the problem. We choose the point *L* so that the triangle *ALC* is isosceles (Fig. 7). Since *D* is supposed to be at infinity, the line *KM* is parallel to *AC*. Therefore the triangles *KAC* and *MAC* are equal, and the triangle *ANC* is also isosceles. The triangles *CNL* and *ANL* are therefore equal, and the line *LB* bisects the angle *ALC*. *B* is therefore the middle point of *AC*, and we have the theorem

The harmonic conjugate of the middle point of AC is at infinity.

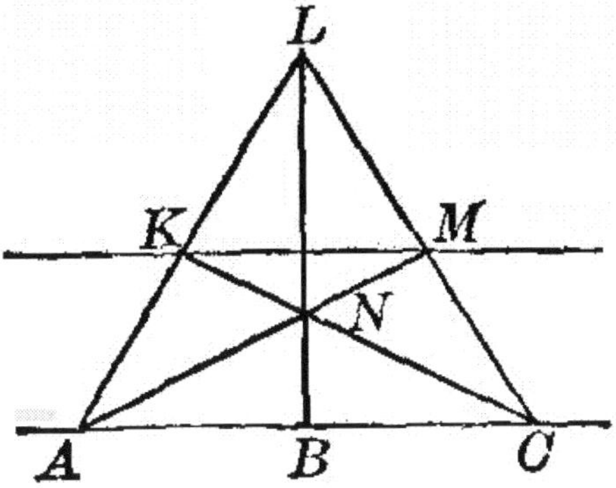

FIG. 7

40. Projective theorems and metrical theorems. Linear construction. This theorem is the connecting link between the general protective theorems which we have been considering so far and the metrical theorems of ordinary geometry. Up to this point we have said nothing about measurements, either of line segments or of angles. Desargues's theorem and the theory of harmonic elements which depends on it have nothing to do with magnitudes at all. Not until the notion of an infinitely distant point is brought in is any mention made of distances or directions. We have been able to make all of our constructions up to this point by means of the straightedge, or ungraduated ruler. A construction made with such an instrument we shall call a *linear* construction. It requires merely that we be able to draw the line joining two points or find the point of intersection of two lines.

41. Parallels and mid-points. It might be thought that drawing a line through a given point parallel to a given line was only a special case of drawing a line joining two points. Indeed, it consists only in drawing a line through the given point and through the "infinitely distant point" on the given line. It must be remembered, however, that the expression "infinitely distant point" must not be taken literally. When we say that two parallel lines meet "at infinity," we really mean that they do not meet at all, and the only reason for using the expression is to avoid tedious statement of exceptions and restrictions to our theorems. We ought therefore to consider the drawing of a line parallel to a given line as a different accomplishment from the drawing of the line joining two given points. It is a remarkable consequence of the last theorem that a parallel to a given line and the mid-point of a given segment are equivalent data. For the construction is reversible, and if we are given the middle point of a given segment, we can construct *linearly* a line parallel to that segment. Thus, given that B is the middle point of AC, we may draw any two lines through A, and any line through B cutting them in points N and L. Join N and L to C and get the points K and M on the two lines through A. Then KM is parallel to AC. *The bisection of a given segment and the drawing of a line parallel to the segment are equivalent data when linear construction is used.*

42. It is not difficult to give a linear construction for the problem to divide a given segment into n equal parts, given only a parallel to the segment. This is simple enough when n is a power of *2*. For any other number, such as *29*, divide any segment on the line parallel to AC into *32* equal parts, by a repetition of the process just described. Take *29* of these, and join the first to A and the last to C. Let these joining lines meet in S. Join S to all the other points. Other problems, of a similar sort, are given at the end of the chapter.

43. Numerical relations. Since three points, given in order, are sufficient to determine a fourth, as explained above, it ought to be possible to reproduce the process numerically in view of the one-to-one correspondence which exists between points on a line and numbers; a correspondence which, to be sure, we have not established here, but which is discussed in any treatise on the theory of point sets. We proceed to discover what relation between four numbers corresponds to the harmonic relation between four points.

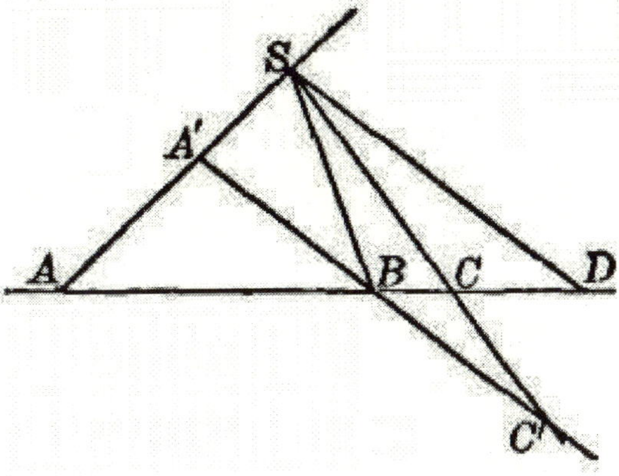

FIG. 8

44. Let A, B, C, D be four harmonic points (Fig. 8), and let SA, SB, SC, SD be four harmonic lines. Assume a line drawn through B parallel to SD, meeting SA in A' and SC in C'. Then A', B', C', and the infinitely distant point on $A'C'$ are four harmonic points, and therefore B is the middle point of the segment $A'C'$. Then, since the triangle DAS is similar to the triangle BAA', we may write the proportion

$$AB : AD = BA' : SD.$$

Also, from the similar triangles DSC and BCC', we have

$$CD : CB = SD : B'C.$$

From these two proportions we have, remembering that $BA' = BC'$,

$$\frac{AB \cdot CD}{AD \cdot CB} = -1,$$

the minus sign being given to the ratio on account of the fact that A and C are always separated from B and D, so that one or three of the segments AB, CD, AD, CB must be negative.

45. Writing the last equation in the form

$$CB : AB = -CD : AD,$$

and using the fundamental relation connecting three points on a line,

$$PR + RQ = PQ,$$

which holds for all positions of the three points if account be taken of the sign of the segments, the last proportion may be written

$$(CB - BA) : AB = -(CA - DA) : AD,$$

or

$$(AB - AC) : AB = (AC - AD) : AD;$$

so that AB, AC, and AD are three quantities in hamonic progression, since the difference between the first and second is to the first as the difference between the second and third is to the third. Also, from this last proportion comes the familiar relation

$$\frac{2}{AC} = \frac{1}{AB} + \frac{1}{AD},$$

which is convenient for the computation of the distance AD when AB and AC are given numerically.

46. Anharmonic ratio. The corresponding relations between the trigonometric functions of the angles determined by four harmonic lines are not difficult to obtain, but as we shall not need them in building up the theory of projective geometry, we will not discuss them here. Students who have a slight acquaintance with trigonometry may read in a later chapter (§ 161) a development of the theory of a more general relation, called the *anharmonic ratio*, or *cross ratio*, which connects any four points on a line.

PROBLEMS

1. Draw through a given point a line which shall pass through the inaccessible point of intersection of two given lines. The following construction may be made to depend upon Desargues's theorem: Through the given point P draw any two rays cutting the two lines in the points AB' and $A'B$, A, B, lying on one of the given lines and A', B', on the other. Join AA' and BB', and find their point of intersection S. Through S draw any other ray, cutting the given lines in CC'. Join BC' and $B'C$, and obtain their point of intersection Q. PQ is the desired line. Justify this construction.

2. To draw through a given point P a line which shall meet two given lines in points A and B, equally distant from P. Justify the following construction: Join P to the point S of intersection of the two given lines. Construct the fourth harmonic of PS with respect to the two given lines. Draw through P a line parallel to this line. This is the required line.

3. Given a parallelogram in the same plane with a given segment AC, to construct linearly the middle point of AC.

4. Given four harmonic lines, of which one pair are at right angles to each other, show that the other pair make equal angles with them. This is a theorem of which frequent use will be made.

5. Given the middle point of a line segment, to draw a line parallel to the segment and passing through a given point.

6. A line is drawn cutting the sides of a triangle ABC in the points A', B', C' the point A' lying on the side BC, etc. The harmonic conjugate of A' with respect to B and C is then constructed and called A''. Similarly, B'' and C'' are constructed. Show that $A''B''C''$ lie on a straight line. Find other sets of three points on a line in the figure. Find also sets of three lines through a point.

CHAPTER III
COMBINATION OF TWO PROJECTIVELY RELATED FUNDAMENTAL FORMS

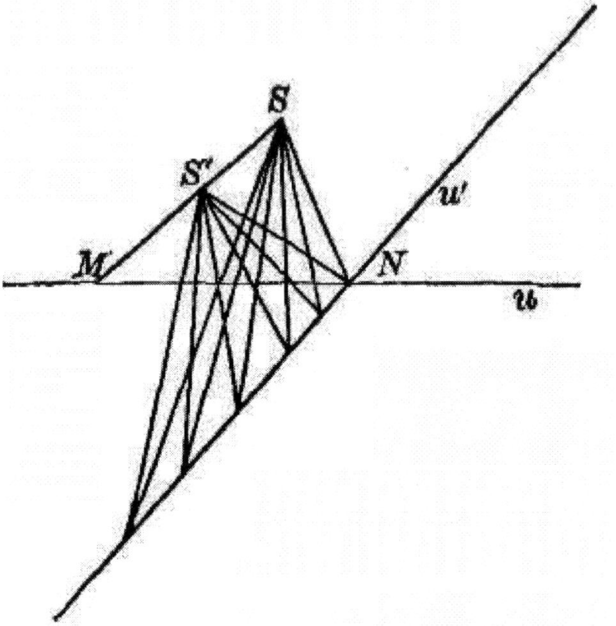

FIG. 9

47. Superposed fundamental forms. Self-corresponding elements. We have seen (§ 37) that two projective point-rows may be superposed upon the same straight line. This happens, for example, when two pencils which are projective to each other are cut across by a straight line. It is also possible for two projective pencils to have the same center. This happens, for example, when two projective point-rows are projected to the same point. Similarly, two projective axial pencils may have the same axis. We examine now the possibility of two forms related in this way, having an element or elements that correspond to themselves. We have seen, indeed, that if *B* and *D* are harmonic conjugates with respect to *A* and *C*, then the point-row described by *B* is projective to the point-row described by *D*, and that *A* and *C* are self-corresponding points. Consider more generally the case of two pencils perspective to each other with axis of perspectivity u' (Fig. 9). Cut across them by a line u. We get thus two projective point-rows superposed on the same line u, and a moment's reflection serves to show that the point N of intersection u and u' corresponds to itself in the two point-rows. Also, the point M, where u intersects the line joining the centers of the two pencils, is seen to correspond to itself. It is thus possible for two projective point-rows, superposed upon the same line, to have two self-corresponding points. Clearly M and N may fall together if the line joining the centers of the pencils happens to pass through the point of intersection of the lines u and u'.

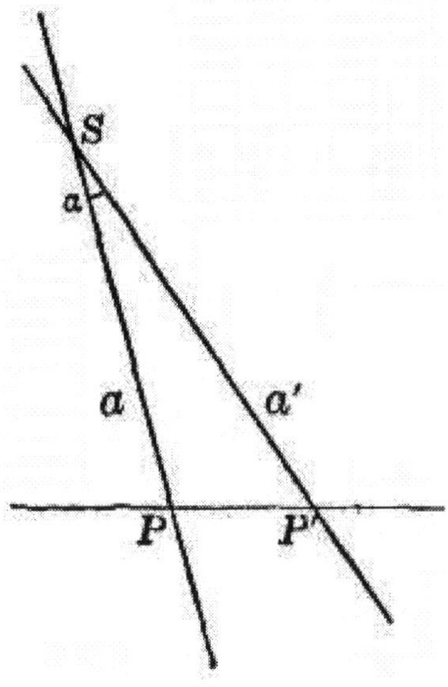

FIG. 10

48. We may also give an illustration of a case where two superposed projective point-rows have no self-corresponding points at all. Thus we may take two lines revolving about a fixed point S and always making the same angle a with each other (Fig. 10). They will cut out on any line u in the plane two point-rows which are easily seen to be projective. For, given any four rays SP which are harmonic, the four corresponding rays SP' must also be harmonic, since they make the same angles with each other. Four harmonic points P correspond, therefore, to four harmonic points P'. It is clear, however, that no point P can coincide with its corresponding point P', for in that case the lines PS and $P'S$ would coincide, which is impossible if the angle between them is to be constant.

49. Fundamental theorem. Postulate of continuity. We have thus shown that two projective point-rows, superposed one on the other, may have two points, one point, or no point at all corresponding to themselves. We proceed to show that

If two projective point-rows, superposed upon the same straight line, have more than two self-corresponding points, they must have an infinite number, and every point corresponds to itself; that is, the two point-rows are not essentially distinct.

If three points, A, B, and C, are self-corresponding, then the harmonic conjugate D of B with respect to A and C must also correspond to itself. For four harmonic points must always correspond to four harmonic points. In the same way the harmonic conjugate of D with respect to B and C must correspond to itself. Combining new points with old in this way, we may obtain as many self-corresponding points as we wish. We show further that every point on the line is the limiting point of a finite or infinite sequence of self-corresponding points. Thus, let a point P lie between A and B. Construct now D, the fourth harmonic of C with respect to A and B. D may coincide with P, in which case the sequence is closed; otherwise P lies in the stretch AD or in the stretch DB. If it lies in the stretch DB, construct the fourth harmonic of C with respect to D and B. This point D' may coincide with P, in which case, as before, the sequence is closed. If P lies in the stretch DD', we construct the fourth harmonic of C with respect to DD', etc. In each step the region in which P lies is diminished, and the process may be continued until two self-corresponding points are obtained on either side of P, and at distances from it arbitrarily small.

We now assume, explicitly, the fundamental postulate that the correspondence is *continuous*, that is, that *the distance between two points in one point-row may be made arbitrarily small by sufficiently diminishing the distance between the corresponding points in the other.* Suppose now that P is not a self-corresponding point, but corresponds to a point P' at a fixed distance d from P. As noted above, we can find self-corresponding points arbitrarily close to P, and it appears, then, that we can take a point D as close to P as we wish, and yet the distance between the corresponding points D' and P' approaches d as a limit, and not zero, which contradicts the postulate of continuity.

50. It follows also that two projective pencils which have the same center may have no more than two self-corresponding rays, unless the pencils are identical. For if we cut across them by a line, we obtain two projective point-rows superposed on the same straight line, which may have no more than two self-corresponding points. The same considerations apply to two projective axial pencils which have the same axis.

51. Projective point-rows having a self-corresponding point in common. Consider now two projective point-rows lying on different lines in the same plane. Their common point may or may not be a self-corresponding point. If the two point-rows are perspectively related, then their common point is evidently a self-corresponding point. The converse is also true, and we have the very important theorem:

52. *If in two protective point-rows, the point of intersection corresponds to itself, then the point-rows are in perspective position.*

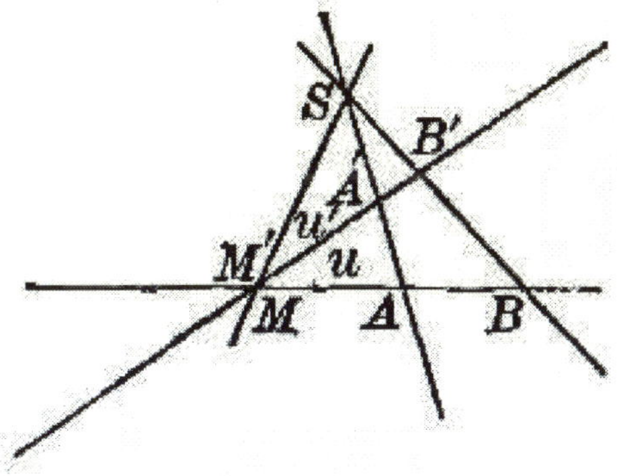

FIG. 11

Let the two point-rows be u and u' (Fig. 11). Let A and A', B and B', be corresponding points, and let also the point M of intersection of u and u' correspond to itself. Let AA' and BB' meet in the point S. Take S as the center of two pencils, one perspective to u and the other perspective to u'. In these two pencils SA coincides with its corresponding ray SA', SB with its corresponding ray SB', and SM with its corresponding ray SM'. The two pencils are thus identical, by the preceding theorem, and any ray SD must coincide with its corresponding ray SD'. Corresponding points of u and u', therefore, all lie on lines through the point S.

53. An entirely similar discussion shows that

If in two projective pencils the line joining their centers is a self-corresponding ray, then the two pencils are perspectively related.

54. A similar theorem may be stated for two axial pencils of which the axes intersect. Very frequent use will be made of these fundamental theorems.

55. Point-row of the second order. The question naturally arises, What is the locus of points of intersection of corresponding rays of two projective pencils which are not in perspective position? This locus, which will be discussed in detail in subsequent chapters, is easily seen to have at most two points in common with any line in the plane, and on account of this fundamental property will be called a *point-row of the second order*. For any line u in the plane of the two pencils will be cut by them in two projective point-rows which have at most two self-corresponding points. Such a self-corresponding point is clearly a point of intersection of corresponding rays of the two pencils.

56. This locus degenerates in the case of two perspective pencils to a pair of straight lines, one of which is the axis of perspectivity and the other the common ray, any point of which may be considered as the point of intersection of corresponding rays of the two pencils.

57. Pencils of rays of the second order. Similar investigations may be made concerning the system of lines joining corresponding points of two projective point-rows. If we project the point-rows to any point in the plane, we obtain two projective pencils having the same center. At most two pairs of self-corresponding rays may present themselves. Such a ray is clearly a line joining two corresponding points in the two point-rows. The result may be stated as follows: *The system of rays joining corresponding points in two protective point-rows has at most two rays in common with any pencil in the plane.* For that reason the system of rays is called *a pencil of rays of the second order.*

58. In the case of two perspective point-rows this system of rays degenerates into two pencils of rays of the first order, one of which has its center at the center of perspectivity of the two point-rows, and the other at the intersection of the two point-rows, any ray through which may be considered as joining two corresponding points of the two point-rows.

59. Cone of the second order. The corresponding theorems in space may easily be obtained by joining the points and lines considered in the plane theorems to a point S in space. Two projective pencils give rise to two projective axial pencils with axes intersecting. Corresponding planes meet in lines which all pass through S and through the points on a point-row of the second order generated by the two pencils of rays. They are thus generating lines of a *cone of the second order*, or *quadric cone*, so called because every plane in space not passing through S cuts it in a point-row of the second order, and every line also cuts it in at most two points. If, again, we project two point-rows to a point S in space, we obtain two pencils of rays with a common center but lying in different planes. Corresponding lines of these pencils determine planes which are the projections to S of the lines which join the corresponding points of the two point-rows. At most two such planes may pass through any ray through S. It is called *a pencil of planes of the second order.*

PROBLEMS

1. A man *A* moves along a straight road *u*, and another man *B* moves along the same road and walks so as always to keep sight of *A* in a small mirror *M* at the side of the road. How many times will they come together, *A* moving always in the same direction along the road?

2. How many times would the two men in the first problem see each other in two mirrors *M* and *N* as they walk along the road as before? (The planes of the two mirrors are not necessarily parallel to *u*.)

3. As A moves along *u*, trace the path of B so that the two men may always see each other in the two mirrors.

4. Two boys walk along two paths *u* and *u'* each holding a string which they keep stretched tightly between them. They both move at constant but different rates of speed, letting out the string or drawing it in as they walk. How many times will the line of the string pass over any given point in the plane of the paths?

5. Trace the lines of the string when the two boys move at the same rate of speed in the two paths but do not start at the same time from the point where the two paths intersect.

6. A ship is sailing on a straight course and keeps a gun trained on a point on the shore. Show that a line at right angles to the direction of the gun at its muzzle will pass through any point in the plane twice or not at all. (Consider the point-row at infinity cut out by a line through the point on the shore at right angles to the direction of the gun.)

7. Two lines *u* and *u'* revolve about two points *U* and *U'* respectively in the same plane. They go in the same direction and at the same rate of speed, but one has an angle a the start of the other. Show that they generate a point-row of the second order.

8. Discuss the question given in the last problem when the two lines revolve in opposite directions. Can you recognize the locus?

CHAPTER IV
POINT-ROWS OF THE SECOND ORDER

60. Point-row of the second order defined. We have seen that two fundamental forms in one-to-one correspondence may sometimes generate a form of higher order. Thus, two point-rows (§ 55) generate a system of rays of the second order, and two pencils of rays (§ 57), a system of points of the second order. As a system of points is more familiar to most students of geometry than a system of lines, we study first the point-row of the second order.

61. Tangent line. We have shown in the last chapter (§ 55) that the locus of intersection of corresponding rays of two projective pencils is a point-row of the second order; that is, it has at most two points in common with any line in the plane. It is clear, first of all, that the centers of the pencils are points of the locus; for to the line SS', considered as a ray of S, must correspond some ray of S' which meets it in S'. S', and by the same argument S, is then a point where corresponding rays meet. Any ray through S will meet it in one point besides S, namely, the point P where it meets its corresponding ray. Now, by choosing the ray through S sufficiently close to the ray SS', the point P may be made to approach arbitrarily close to S', and the ray $S'P$ may be made to differ in position from the tangent line at S' by as little as we please. We have, then, the important theorem

The ray at S' which corresponds to the common ray SS' is tangent to the locus at S'.

In the same manner the tangent at S may be constructed.

62. Determination of the locus. We now show that *it is possible to assign arbitrarily the position of three points, A, B, and C, on the locus (besides the points S and S'); but, these three points being chosen, the locus is completely determined.*

63. This statement is equivalent to the following:

Given three pairs of corresponding rays in two projective pencils, it is possible to find a ray of one which corresponds to any ray of the other.

64. We proceed, then, to the solution of the fundamental

PROBLEM: *Given three pairs of rays, aa′, bb′, and cc′, of two protective pencils, S and S′, to find the ray d′ of S′ which corresponds to any ray d of S.*

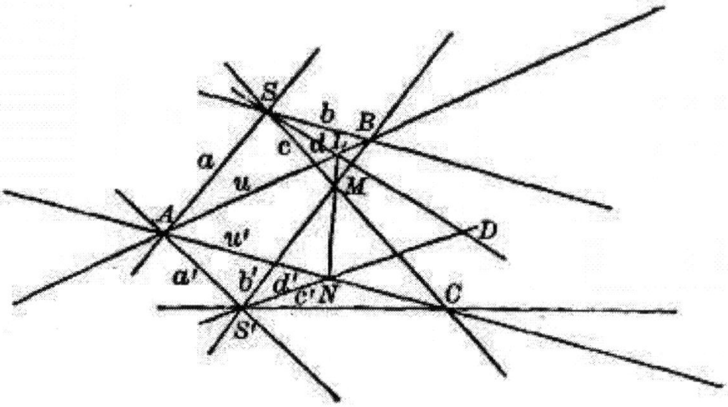

FIG. 12

Call *A* the intersection of *aa′*, *B* the intersection of *bb′*, and *C* the intersection of *cc′* (Fig. 12). Join *AB* by the line *u*, and *AC* by the line *u′*. Consider *u* as a point-row perspective to *S*, and *u′* as a point-row perspective to *S′*. *u* and *u′* are projectively related to each other, since *S* and *S′* are, by hypothesis, so related. But their point of intersection *A* is a self-corresponding point, since *a* and *a′* were supposed to be corresponding rays. It follows (§ 52) that *u* and *u′* are in perspective position, and that lines through corresponding points all pass through a point *M*, the center of perspectivity, the position of which will be determined by any two such lines. But the intersection of *a* with *u* and the intersection of *c′* with *u′* are corresponding points on *u* and *u′*, and the line joining them is clearly *c* itself. Similarly, *b′* joins two corresponding points on *u* and *u′*, and so the center *M* of perspectivity of *u* and *u′* is the intersection of *c* and *b′*. To find *d′* in *S′* corresponding to a given line *d* of *S* we note the point *L* where *d* meets *u*. Join *L* to *M* and get the point *N* where this line meets u'. L and N are corresponding points on u and u', and d' must therefore pass through N. The intersection P of d and d' is thus another point on the locus. In the same manner any number of other points may be obtained.

65. The lines u and u' might have been drawn in any direction through A (avoiding, of course, the line a for u and the line a' for u'), and the center of perspectivity M would be easily obtainable; but the above construction furnishes a simple and instructive figure. An equally simple one is obtained by taking a' for u and a for u'.

66. Lines joining four points of the locus to a fifth. Suppose that the points S, S', B, C, and D are fixed, and that four points, A, A_1, A_2, and A_3, are taken on the locus at the intersection with it of any four harmonic rays through B. These four harmonic rays give four harmonic points, L, L_1 etc., on the fixed ray SD. These, in turn, project through the fixed point M into four harmonic points, N, N_1 etc., on the fixed line DS'. These last four harmonic points give four harmonic rays CA, CA_1, CA_2, CA_3. Therefore the four points A which project to B in four harmonic rays also project to C in four harmonic rays. But C may be any point on the locus, and so we have the very important theorem,

Four points which are on the locus, and which project to a fifth point of the locus in four harmonic rays, project to any point of the locus in four harmonic rays.

67. The theorem may also be stated thus:

The locus of points from which, four given points are seen along four harmonic rays is a point-row of the second order through them.

68. A further theorem of prime importance also follows:

Any two points on the locus may be taken as the centers of two projective pencils which will generate the locus.

69. Pascal's theorem. The points A, B, C, D, S, and S' may thus be considered as chosen arbitrarily on the locus, and the following remarkable theorem follows at once.

Given six points, 1, 2, 3, 4, 5, 6, on the point-row of the second order, if we call

L the intersection of 12 with 45,

M the intersection of 23 with 56,

N the intersection of 34 with 61,

then L, M, and N are on a straight line.

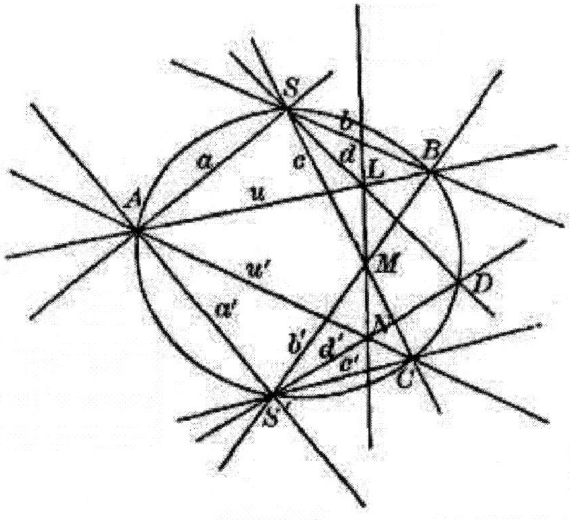

FIG. 13

70. To get the notation to correspond to the figure, we may take (Fig. 13) $A = 1$, $B = 2$, $S' = 3$, $D = 4$, $S = 5$, and $C = 6$. If we make $A = 1$, $C=2$, $S=3$, $D = 4$, $S'=5$, and. $B = 6$, the points L and N are interchanged, but the line is left unchanged. It is clear that one point may be named arbitrarily and the other five named in $5! = 120$ different ways, but since, as we have seen, two different assignments of names give the same line, it follows that there cannot be more than 60 different lines LMN obtained in this way from a given set of six points. As a matter of fact, the number obtained in this way is in general *60*. The above theorem, which is of cardinal importance in the theory of the point-row of the second order, is due to Pascal and was discovered by him at the age of sixteen. It is, no doubt, the most important contribution to the theory of these loci since the days of Apollonius. If the six points be called the vertices of a hexagon inscribed in the curve, then the sides 12 and 45 may be appropriately called a pair of opposite sides. Pascal's theorem, then, may be stated as follows:

The three pairs of opposite sides of a hexagon inscribed in a point-row of the second order meet in three points on a line.

71. Harmonic points on a point-row of the second order. Before proceeding to develop the consequences of this theorem, we note another result of the utmost importance for the higher developments of pure geometry, which follows from the fact that if four points on the locus project to a fifth in four harmonic rays, they will project to any point of the locus in four harmonic rays. It is natural to speak of four such points as four harmonic points on the locus, and to use this notion to define projective correspondence between point-rows of the second order, or between a point-row of the second order and any fundamental form of the first order. Thus, in particular, the point-row of the second order, σ, is said to be *perspectively related* to the pencil S when every ray on S goes through the point on σ which corresponds to it.

72. Determination of the locus. It is now clear that five points, arbitrarily chosen in the plane, are sufficient to determine a point-row of the second order through them. Two of the points may be taken as centers of two projective pencils, and the three others will determine three pairs of corresponding rays of the pencils, and therefore all pairs. If four points of the locus are given, together with the tangent at one of them, the locus is likewise completely determined. For if the point at which the tangent is given be taken as the center S of one pencil, and any other of the points for S', then, besides the two pairs of corresponding rays determined by the remaining two points, we have one more pair, consisting of the tangent at S and the ray SS'. Similarly, the curve is determined by three points and the tangents at two of them.

73. Circles and conics as point-rows of the second order. It is not difficult to see that a circle is a point-row of the second order. Indeed, take any point *S* on the circle and draw four harmonic rays through it. They will cut the circle in four points, which will project to any other point of the curve in four harmonic rays; for, by the theorem concerning the angles inscribed in a circle, the angles involved in the second set of four lines are the same as those in the first set. If, moreover, we project the figure to any point in space, we shall get a cone, standing on a circular base, generated by two projective axial pencils which are the projections of the pencils at *S* and *S'*. Cut across, now, by any plane, and we get a conic section which is thus exhibited as the locus of intersection of two projective pencils. It thus appears that a conic section is a point-row of the second order. It will later appear that a point-row of the second order is a conic section. In the future, therefore, we shall refer to a point-row of the second order as a conic.

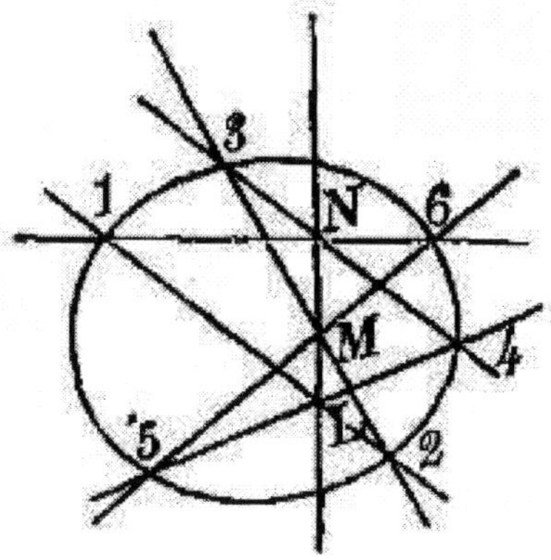

FIG. 14

74. Conic through five points. Pascal's theorem furnishes an elegant solution of the problem of drawing a conic through five given points. To construct a sixth point on the conic, draw through the point numbered 1 an arbitrary line (Fig. 14), and let the desired point 6 be the second point of intersection of this line with the conic. The point *L* = *12-45* is obtainable at once; also the point *N* = *34-61*. But *L* and *N* determine Pascal's line, and the intersection of 23 with 56 must be on this line. Intersect, then, the line *LN* with 23 and obtain the point *M*. Join *M* to 5 and intersect with 61 for the desired point 6.

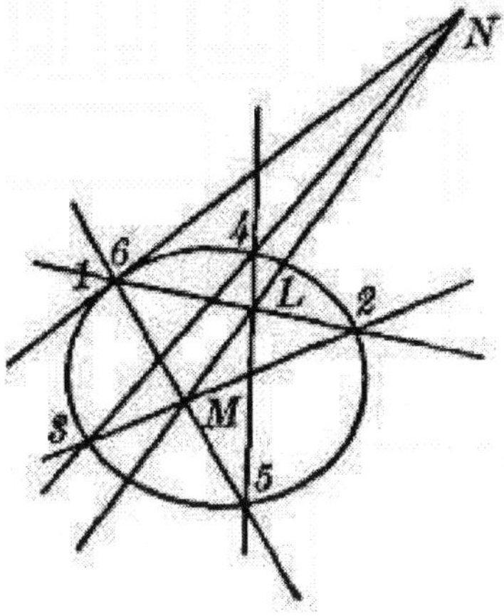

FIG. 15

75. Tangent to a conic. If two points of Pascal's hexagon approach coincidence, then the line joining them approaches as a limiting position the tangent line at that point. Pascal's theorem thus affords a ready method of drawing the tangent line to a conic at a given point. If the conic is determined by the points 1, 2, 3, 4, 5 (Fig. 15), and it is desired to draw the tangent at the point 1, we may call that point 1, 6. The points *L* and *M* are obtained as usual, and the intersection of 34 with *LM* gives *N*. Join *N* to the point 1 for the desired tangent at that point.

76. Inscribed quadrangle. Two pairs of vertices may coalesce, giving an inscribed quadrangle. Pascal's theorem gives for this case the very important theorem

Two pairs of opposite sides of any quadrangle inscribed in a conic meet on a straight line, upon which line also intersect the two pairs of tangents at the opposite vertices.

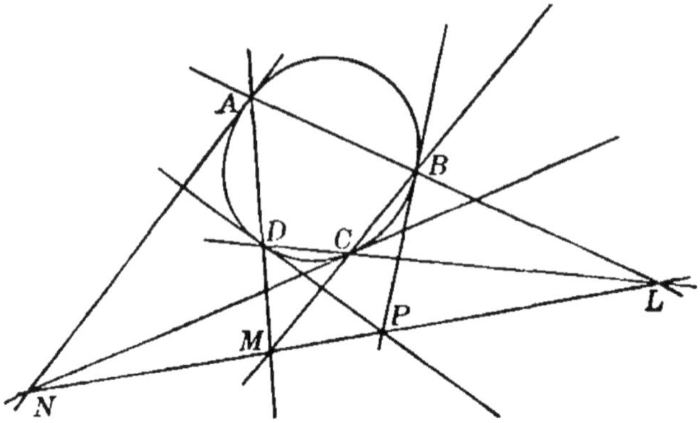

FIG. 16

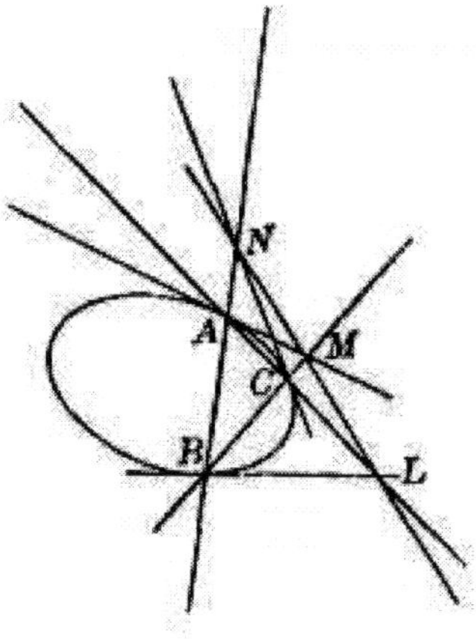

FIG. 17

For let the vertices be *A*, *B*, *C*, and *D*, and call the vertex *A* the point 1, 6; *B*, the point 2; *C*, the point 3, 4; and *D*, the point 5 (Fig. 16). Pascal's theorem then indicates that *L* = *AB-CD*, *M* = *AD-BC*, and *N*, which is the intersection of the tangents at *A* and *C*, are all on a straight line *u*. But if we were to call *A* the point 2, *B* the point 6, 1, *C* the point 5, and *D* the point 4, 3, then the intersection *P* of the tangents at *B* and *D* are also on this same line *u*. Thus *L*, *M*, *N*, and *P* are four points on a straight line. The consequences of this theorem are so numerous and important that we shall devote a separate chapter to them.

77. Inscribed triangle. Finally, three of the vertices of the hexagon may coalesce, giving a triangle inscribed in a conic. Pascal's theorem then reads as follows (Fig. 17) for this case:

The three tangents at the vertices of a triangle inscribed in a conic meet the opposite sides in three points on a straight line.

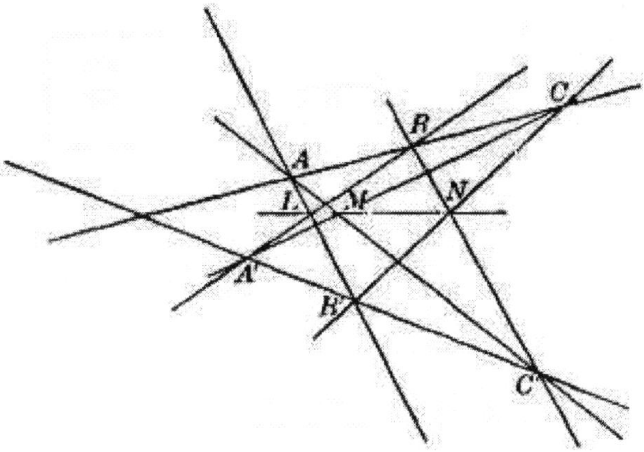

FIG. 18

78. Degenerate conic. If we apply Pascal's theorem to a degenerate conic made up of a pair of straight lines, we get the following theorem (Fig. 18):

If three points, A, B, C, are chosen on one line, and three points, A', B', C', are chosen on another, then the three points L = AB'-A'B, M = BC'-B'C, N = CA'-C'A are all on a straight line.

PROBLEMS

1. In Fig. 12, select different lines u and trace the locus of the center of perspectivity M of the lines u and u'.

2. Given four points, A, B, C, D, in the plane, construct a fifth point P such that the lines PA, PB, PC, PD shall be four harmonic lines.

Suggestion. Draw a line a through the point A such that the four lines a, AB, AC, AD are harmonic. Construct now a conic through A, B, C, and D having a for a tangent at A.

3. Where are all the points P, as determined in the preceding question, to be found?

4. Select any five points in the plane and draw the tangent to the conic through them at each of the five points.

5. Given four points on the conic, and the tangent at one of them, to construct the conic. ("To construct the conic" means here to construct as many other points as may be desired.)

6. Given three points on the conic, and the tangent at two of them, to construct the conic.

7. Given five points, two of which are at infinity in different directions, to construct the conic. (In this, and in the following examples, the student is supposed to be able to draw a line parallel to a given line.)

8. Given four points on a conic (two of which are at infinity and two in the finite part of the plane), together with the tangent at one of the finite points, to construct the conic.

9. The tangents to a curve at its infinitely distant points are called its *asymptotes* if they pass through a finite part of the plane. Given the asymptotes and a finite point of a conic, to construct the conic.

10. Given an asymptote and three finite points on the conic, to determine the conic.

11. Given four points, one of which is at infinity, and given also that the line at infinity is a tangent line, to construct the conic.

CHAPTER V
PENCILS OF RAYS OF THE SECOND ORDER

79. Pencil of rays of the second order defined. If the corresponding points of two projective point-rows be joined by straight lines, a system of lines is obtained which is called a pencil of rays of the second order. This name arises from the fact, easily shown (§ 57), that at most two lines of the system may pass through any arbitrary point in the plane. For if through any point there should pass three lines of the system, then this point might be taken as the center of two projective pencils, one projecting one point-row and the other projecting the other. Since, now, these pencils have three rays of one coincident with the corresponding rays of the other, the two are identical and the two point-rows are in perspective position, which was not supposed.

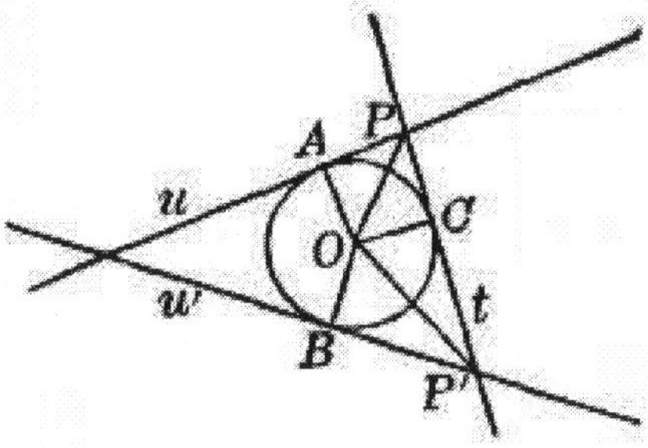

FIG. 19

80. Tangents to a circle. To get a clear notion of this system of lines, we may first show that the tangents to a circle form a system of this kind. For take any two tangents, u and u', to a circle, and let A and B be the points of contact (Fig. 19). Let now t be any third tangent with point of contact at C and meeting u and u' in P and P' respectively. Join A, B, P, P', and C to O, the center of the circle. Tangents from any point to a circle are equal, and therefore the triangles POA and POC are equal, as also are the triangles $P'OB$ and $P'OC$. Therefore the angle POP' is constant, being equal to half the constant angle $AOC + COB$. This being true, if we take any four harmonic points, P_1, P_2, P_3, P_4, on the line u, they will project to O in four harmonic lines, and the tangents to the circle from these four points will meet *u'* in four harmonic points, P'_1, P'_2, P'_3, P'_4, because the lines from these points to *O* inclose the same angles as the lines from the points P_1, P_2, P_3, P_4 on *u*. The point-row on *u* is therefore projective to the point-row on *u'*. Thus the tangents to a circle are seen to join corresponding points on two projective point-rows, and so, according to the definition, form a pencil of rays of the second order.

81. Tangents to a conic. If now this figure be projected to a point outside the plane of the circle, and any section of the resulting cone be made by a plane, we can easily see that the system of rays tangent to any conic section is a pencil of rays of the second order. The converse is also true, as we shall see later, and a pencil of rays of the second order is also a set of lines tangent to a conic section.

82. The point-rows *u* and *u'* are, themselves, lines of the system, for to the common point of the two point-rows, considered as a point of *u*, must correspond some point of *u'*, and the line joining these two corresponding points is clearly *u'* itself. Similarly for the line *u*.

83. Determination of the pencil. We now show that *it is possible to assign arbitrarily three lines, a, b, and c, of the system (besides the lines u and u'); but if these three lines are chosen, the system is completely determined.*

This statement is equivalent to the following:

Given three pairs of corresponding points in two projective point-rows, it is possible to find a point in one which corresponds to any point of the other.

We proceed, then, to the solution of the fundamental

PROBLEM. *Given three pairs of points, AA', BB', and CC', of two projective point-rows u and u', to find the point D' of u' which corresponds to any given point D of u.*

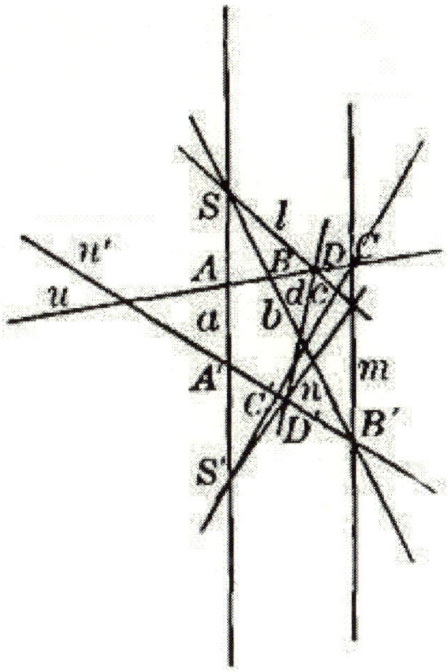

FIG. 20

On the line a, joining A and A', take two points, S and S', as centers of pencils perspective to u and u' respectively (Fig. 20). The figure will be much simplified if we take S on BB' and S' on CC'. SA and $S'A'$ are corresponding rays of S and S', and the two pencils are therefore in perspective position. It is not difficult to see that the axis of perspectivity m is the line joining B' and C. Given any point D on u, to find the corresponding point D' on u' we proceed as follows: Join D to S and note where the joining line meets m. Join this point to S'. This last line meets u' in the desired point D'.

We have now in this figure six lines of the system, a, b, c, d, u, and u'. Fix now the position of u, u', b, c, and d, and take four lines of the system, a_1, a_2, a_3, a_4, which meet b in four harmonic points. These points project to D, giving four harmonic points on m. These again project to D', giving four harmonic points on c. It is thus clear that the rays a_1, a_2, a_3, a_4 cut out two projective point-rows on any two lines of the system. Thus u and u' are not special rays, and any two rays of the system will serve as the point-rows to generate the system of lines.

84. Brianchon's theorem. From the figure also appears a fundamental theorem due to Brianchon:

If 1, 2, 3, 4, 5, 6 are any six rays of a pencil of the second order, then the lines l = (12, 45), m = (23, 56), n = (34, 61) all pass through a point.

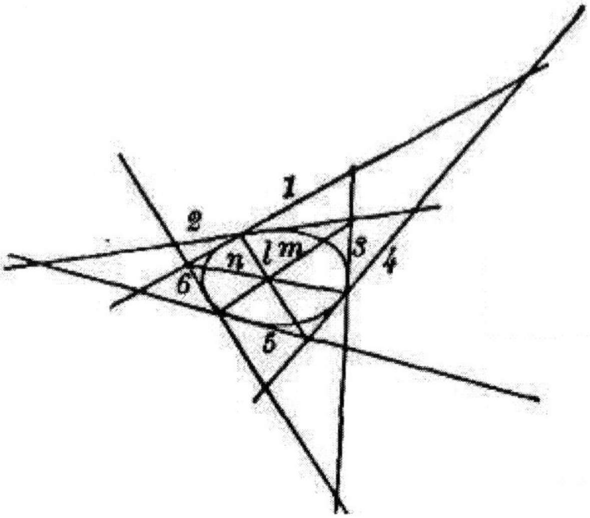

FIG. 21

85. To make the notation fit the figure (Fig. 21), make $a=1$, $b = 2$, $u' = 3$, $d = 4$, $u = 5$, $c = 6$; or, interchanging two of the lines, $a = 1$, $c = 2$, $u = 3$, $d = 4$, $u' = 5$, $b = 6$. Thus, by different namings of the lines, it appears that not more than 60 different *Brianchon points* are possible. If we call 12 and 45 opposite vertices of a circumscribed hexagon, then Brianchon's theorem may be stated as follows:

The three lines joining the three pairs of opposite vertices of a hexagon circumscribed about a conic meet in a point.

86. Construction of the pencil by Brianchon's theorem. Brianchon's theorem furnishes a ready method of determining a sixth line of the pencil of rays of the second order when five are given. Thus, select a point in line

1 and suppose that line 6 is to pass through it. Then *l* = *(12, 45)*, *n* = *(34, 61)*, and the line *m* = *(23, 56)* must pass through *(l, n)*. Then *(23, ln)* meets 5 in a point of the required sixth line.

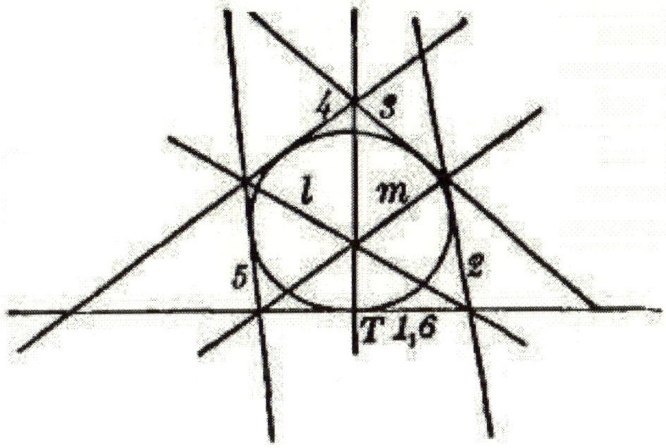

FIG. 22

87. Point of contact of a tangent to a conic. If the line 2 approach as a limiting position the line 1, then the intersection *(1, 2)* approaches as a limiting position the point of contact of 1 with the conic. This suggests an easy way to construct the point of contact of any tangent with the conic. Thus (Fig. 22), given the lines 1, 2, 3, 4, 5 to construct the point of contact of *1=6*. Draw *l* = *(12,45)*, *m* =*(23,56)*; then *(34, lm)* meets 1 in the required point of contact *T*.

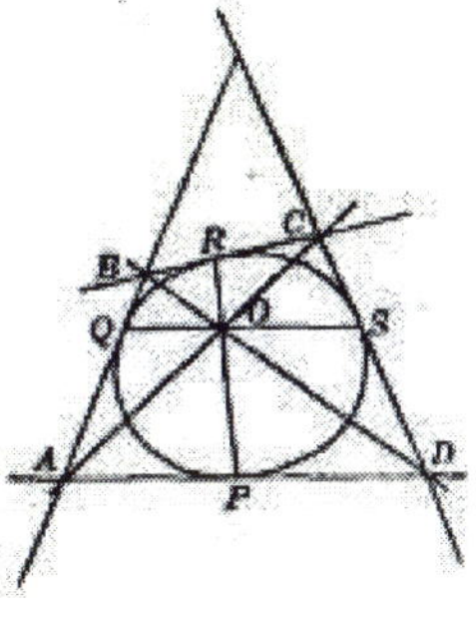

FIG. 23

88. Circumscribed quadrilateral. If two pairs of lines in Brianchon's hexagon coalesce, we have a theorem concerning a quadrilateral circumscribed about a conic. It is easily found to be (Fig. 23)

The four lines joining the two opposite pairs of vertices and the two opposite points of contact of a quadrilateral circumscribed about a conic all meet in a point. The consequences of this theorem will be deduced later.

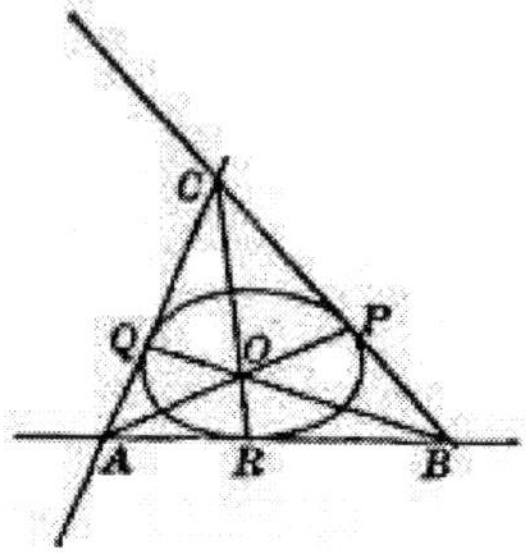

FIG. 24

89. Circumscribed triangle. The hexagon may further degenerate into a triangle, giving the theorem (Fig. 24) *The lines joining the vertices to the points of contact of the opposite sides of a triangle circumscribed about a conic all meet in a point.*

90. Brianchon's theorem may also be used to solve the following problems:

Given four tangents and the point of contact on any one of them, to construct other tangents to a conic. Given three tangents and the points of contact of any two of them, to construct other tangents to a conic.

91. Harmonic tangents. We have seen that a variable tangent cuts out on any two fixed tangents projective point-rows. It follows that if four tangents cut a fifth in four harmonic points, they must cut every tangent in four harmonic points. It is possible, therefore, to make the following definition:

Four tangents to a conic are said to be harmonic when they meet every other tangent in four harmonic points.

92. Projectivity and perspectivity. This definition suggests the possibility of defining a projective correspondence between the elements of a pencil of rays of the second order and the elements of any form heretofore discussed. In particular, the points on a tangent are said to be *perspectively related* to the tangents of a conic when each point lies on the tangent which corresponds to it. These notions are of importance in the higher developments of the subject.

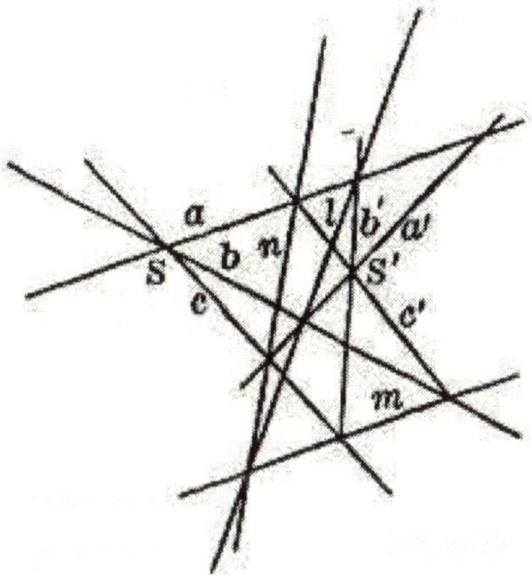

FIG. 25

93. Brianchon's theorem may also be applied to a degenerate conic made up of two points and the lines through them. Thus(Fig. 25),

If a, b, c are three lines through a point S, and a', b', c' are three lines through another point S', then the lines $l = (ab', a'b)$, $m = (bc', b'c)$, and $n = (ca', c'a)$ all meet in a point.

94. Law of duality. The observant student will not have failed to note the remarkable similarity between the theorems of this chapter and those of the preceding. He will have noted that points have replaced lines and lines have replaced points; that points on a curve have been replaced by tangents to a curve; that pencils have been replaced by point-rows, and that a conic considered as made up of a succession of points has been replaced by a conic considered as generated by a moving tangent line. The theory upon which this wonderful *law of duality* is based will be developed in the next chapter.

PROBLEMS

1. Given four lines in the plane, to construct another which shall meet them in four harmonic points.

2. Where are all such lines found?

3. Given any five lines in the plane, construct on each the point of contact with the conic tangent to them all.

4. Given four lines and the point of contact on one, to construct the conic. ("To construct the conic" means here to draw as many other tangents as may be desired.)

5. Given three lines and the point of contact on two of them, to construct the conic.

6. Given four lines and the line at infinity, to construct the conic.

7. Given three lines and the line at infinity, together with the point of contact at infinity, to construct the conic.

8. Given three lines, two of which are asymptotes, to construct the conic.

9. Given five tangents to a conic, to draw a tangent which shall be parallel to any one of them.

10. The lines a, b, c are drawn parallel to each other. The lines a', b', c' are also drawn parallel to each other. Show why the lines $(ab', a'b)$, $(bc', b'c)$, $(ca', c'a)$ meet in a point. (In problems 6 to 10 inclusive, parallel lines are to be drawn.)

CHAPTER VI
POLES AND POLARS

95. Inscribed and circumscribed quadrilaterals. The following theorems have been noted as special cases of Pascal's and Brianchon's theorems:

If a quadrilateral be inscribed in a conic, two pairs of opposite sides and the tangents at opposite vertices intersect in four points, all of which lie on a straight line.

If a quadrilateral be circumscribed about a conic, the lines joining two pairs of opposite vertices and the lines joining two opposite points of contact are four lines which meet in a point.

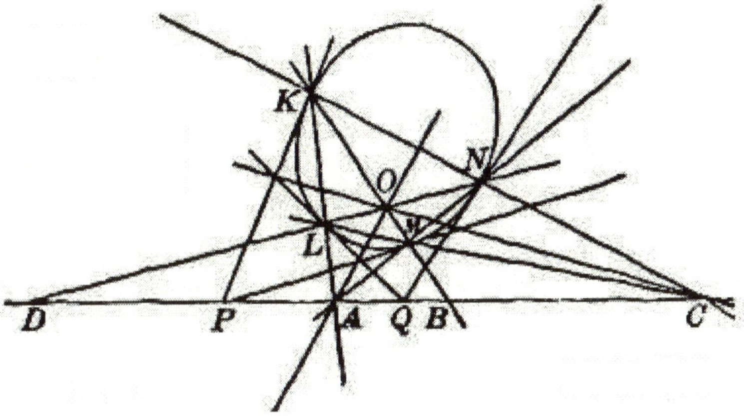

FIG. 26

96. Definition of the polar line of a point. Consider the quadrilateral *K, L, M, N* inscribed in the conic (Fig. 26). It determines the four harmonic points *A, B, C, D* which project from *N* in to the four harmonic points *M, B, K, O*. Now the tangents at *K* and *M* meet in *P*, a point on the line *AB*. The line *AB* is thus determined entirely by the point *O*. For if we draw any line through it, meeting the conic in *K* and *M*, and construct the harmonic conjugate *B* of *O* with respect to *K* and *M*, and also the two tangents at *K* and *M* which meet in the point *P*, then *BP* is the line in question. It thus appears that the line

LON may be any line whatever through *O*; and since *D*, *L*, *O*, *N* are four harmonic points, we may describe the line *AB* as the locus of points which are harmonic conjugates of *O* with respect to the two points where any line through *O* meets the curve.

97. Furthermore, since the tangents at *L* and *N* meet on this same line, it appears as the locus of intersections of pairs of tangents drawn at the extremities of chords through *O*.

98. This important line, which is completely determined by the point *O*, is called the *polar* of *O* with respect to the conic; and the point *O* is called the *pole* of the line with respect to the conic.

99. If a point *B* is on the polar of *O*, then it is harmonically conjugate to *O* with respect to the two intersections *K* and *M* of the line *BC* with the conic. But for the same reason *O* is on the polar of *B*. We have, then, the fundamental theorem

If one point lies on the polar of a second, then the second lies on the polar of the first.

100. Conjugate points and lines. Such a pair of points are said to be *conjugate* with respect to the conic. Similarly, lines are said to be *conjugate* to each other with respect to the conic if one, and consequently each, passes through the pole of the other.

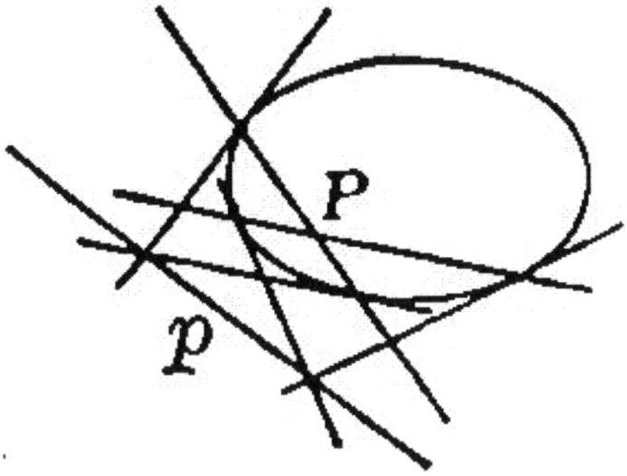

FIG. 27

101. Construction of the polar line of a given point. Given a point *P*, if it is within the conic (that is, if no tangents may be drawn from *P* to the conic), we may construct its polar line by drawing through it any two chords and joining the two points of intersection of the two pairs of tangents at their extremities. If the point *P* is outside the conic, we may draw the two tangents and construct the chord of contact (Fig. 27).

102. Self-polar triangle. In Fig. 26 it is not difficult to see that *AOC* is a *self-polar* triangle, that is, each vertex is the pole of the opposite side. For *B*, *M*, *O*, *K* are four harmonic points, and they project to *C* in four harmonic rays. The line *CO*, therefore, meets the line *AMN* in a point on the polar of *A*, being separated from *A* harmonically by the points *M* and *N*. Similarly, the line *CO* meets *KL* in a point on the polar of *A*, and therefore *CO* is the polar of *A*. Similarly, *OA* is the polar of *C*, and therefore *O* is the pole of *AC*.

103. Pole and polar projectively related. Another very important theorem comes directly from Fig. 26.

As a point A moves along a straight line its polar with respect to a conic revolves about a fixed point and describes a pencil projective to the point-row described by A.

For, fix the points L and N and let the point A move along the line AQ; then the point-row A is projective to the pencil LK, and since K moves along the conic, the pencil LK is projective to the pencil NK, which in turn is projective to the point-row C, which, finally, is projective to the pencil OC, which is the polar of A.

104. Duality. We have, then, in the pole and polar relation a device for setting up a one-to-one correspondence between the points and lines of the plane—a correspondence which may be called projective, because to four harmonic points or lines correspond always four harmonic lines or points. To every figure made up of points and lines will correspond a figure made up of lines and points. To a point-row of the second order, which is a conic considered as a point-locus, corresponds a pencil of rays of the second order, which is a conic considered as a line-locus. The name 'duality' is used to describe this sort of correspondence. It is important to note that the dual relation is subject to the same exceptions as the one-to-one correspondence is, and must not be appealed to in cases where the one-to-one correspondence breaks down. We have seen that there is in Euclidean geometry one and only one ray in a pencil which has no point in a point-row perspective to it for a corresponding point; namely, the line parallel to the line of the point-row. Any theorem, therefore, that involves explicitly the point at infinity is not to be translated into a theorem concerning lines. Further, in the pencil the angle between two lines has nothing to correspond to it in a point-row perspective to the pencil. Any theorem, therefore, that mentions angles is not translatable into another theorem by means of the law of duality. Now we have seen that the notion of the infinitely distant point on a line involves the notion of dividing a segment into any number of equal parts—in other words, of *measuring*. If, therefore, we call any theorem that has to do with the line at infinity or with the measurement of angles a *metrical* theorem, and any other kind a *projective* theorem, we may put the case as follows:

Any projective theorem involves another theorem, dual to it, obtainable by interchanging everywhere the words 'point' and 'line.'

105. Self-dual theorems. The theorems of this chapter will be found, upon examination, to be *self-dual*; that is, no new theorem results from applying the process indicated in the preceding paragraph. It is therefore useless to look for new results from the theorem on the circumscribed quadrilateral derived from Brianchon's, which is itself clearly the dual of Pascal's theorem, and in fact was first discovered by dualization of Pascal's.

106. It should not be inferred from the above discussion that one-to-one correspondences may not be devised that will control certain of the so-called metrical relations. A very important one may be easily found that leaves angles unaltered. The relation called *similarity* leaves ratios between corresponding segments unaltered. The above statements apply only to the particular one-to-one correspondence considered.

PROBLEMS

1. Given a quadrilateral, construct the quadrangle polar to it with respect to a given conic.

2. A point moves along a straight line. Show that its polar lines with respect to two given conics generate a point-row of the second order.

3. Given five points, draw the polar of a point with respect to the conic passing through them, without drawing the conic itself.

4. Given five lines, draw the polar of a point with respect to the conic tangent to them, without drawing the conic itself.

5. Dualize problems 3 and 4.

6. Given four points on the conic, and the tangent at one of them, draw the polar of a given point without drawing the conic. Dualize.

7. A point moves on a conic. Show that its polar line with respect to another conic describes a pencil of rays of the second order.

Suggestion. Replace the given conic by a pair of protective pencils.

8. Show that the poles of the tangents of one conic with respect to another lie on a conic.

9. The polar of a point A with respect to one conic is a, and the pole of a with respect to another conic is A'. Show that as A travels along a line, A' also travels along another line. In general, if A describes a curve of degree n, show that A' describes another curve of the same degree n. (The degree of a curve is the greatest number of points that it may have in common with any line in the plane.)

CHAPTER VII
METRICAL PROPERTIES OF THE CONIC SECTIONS

107. Diameters. Center. After what has been said in the last chapter one would naturally expect to get at the metrical properties of the conic sections by the introduction of the infinite elements in the plane. Entering into the theory of poles and polars with these elements, we have the following definitions:

The polar line of an infinitely distant point is called a *diameter*, and the pole of the infinitely distant line is called the *center*, of the conic.

108. From the harmonic properties of poles and polars,

The center bisects all chords through it (§ 39).

Every diameter passes through the center.

All chords through the same point at infinity (that is, each of a set of parallel chords) are bisected by the diameter which is the polar of that infinitely distant point.

109. Conjugate diameters. We have already defined conjugate lines as lines which pass each through the pole of the other (§ 100).

Any diameter bisects all chords parallel to its conjugate.

The tangents at the extremities of any diameter are parallel, and parallel to the conjugate diameter.

Diameters parallel to the sides of a circumscribed parallelogram are conjugate.

All these theorems are easy exercises for the student.

110. Classification of conics. Conics are classified according to their relation to the infinitely distant line. If a conic has two points in common with the line at infinity, it is called a *hyperbola*; if it has no point in common with the infinitely distant line, it is called an *ellipse*; if it is tangent to the line at infinity, it is called a *parabola*.

111. *In a hyperbola the center is outside the curve* (§ 101), since the two tangents to the curve at the points where it meets the line at infinity determine by their intersection the center. As previously noted, these two tangents are called the *asymptotes* of the curve. The ellipse and the parabola have no asymptotes.

112. *The center of the parabola is at infinity, and therefore all its diameters are parallel,* for the pole of a tangent line is the point of contact.

The locus of the middle points of a series of parallel chords in a parabola is a diameter, and the direction of the line of centers is the same for all series of parallel chords.

The center of an ellipse is within the curve.

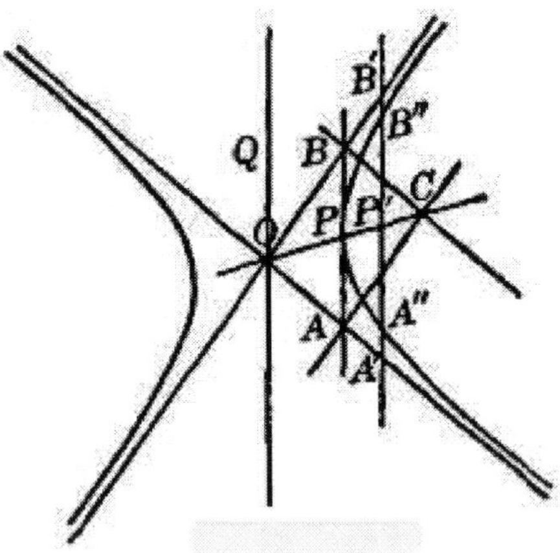

FIG. 28

113. Theorems concerning asymptotes. We derived as a consequence of the theorem of Brianchon (§ 89) the proposition that if a triangle be circumscribed about a conic, the lines joining the vertices to the points of contact of the opposite sides all meet in a point. Take, now, for two of the tangents the asymptotes of a hyperbola, and let any third tangent cut them in A and B (Fig. 28). If, then, O is the intersection of the asymptotes,—and therefore the center of the curve,— then the triangle OAB is circumscribed about the curve. By the theorem just quoted, the line through A parallel to OB, the line through B parallel to OA, and the line OP through the point of contact of the tangent AB all meet in a point C. But $OACB$ is a parallelogram, and $PA = PB$. Therefore

The asymptotes cut off on each tangent a segment which is bisected by the point of contact.

114. If we draw a line OQ parallel to AB, then OP and OQ are conjugate diameters, since OQ is parallel to the tangent at the point where OP meets the curve. Then, since A, P, B, and the point at infinity on AB are four harmonic points, we have the theorem

Conjugate diameters of the hyperbola are harmonic conjugates with respect to the asymptotes.

115. The chord $A''B''$, parallel to the diameter OQ, is bisected at P' by the conjugate diameter OP. If the chord $A''B''$ meet the asymptotes in A', B', then A', P', B', and the point at infinity are four harmonic points, and therefore P' is the middle point of $A'B'$. Therefore $A'A'' = B'B''$ and we have the theorem

The segments cut off on any chord between the hyperbola and its asymptotes are equal.

116. This theorem furnishes a ready means of constructing the hyperbola by points when a point on the curve and the two asymptotes are given.

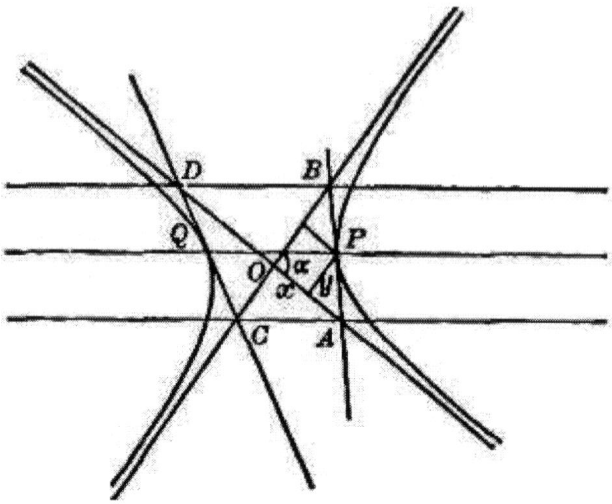

FIG. 29

117. For the circumscribed quadrilateral, Brianchon's theorem gave (§ 88) *The lines joining opposite vertices and the lines joining opposite points of contact are four lines meeting in a point.* Take now for two of the tangents the asymptotes, and let *AB* and *CD* be any other two (Fig. 29). If *B* and *D* are opposite vertices, and also *A* and *C*, then *AC* and *BD* are parallel, and parallel to *PQ*, the line joining the points of contact of *AB* and *CD*, for these are three of the four lines of the theorem just quoted. The fourth is the line at infinity which joins the point of contact of the asymptotes. It is thus seen that the triangles *ABC* and *ADC* are equivalent, and therefore the triangles *AOB* and *COD* are also. The tangent AB may be fixed, and the tangent *CD* chosen arbitrarily; therefore

The triangle formed by any tangent to the hyperbola and the two asymptotes is of constant area.

118. Equation of hyperbola referred to the asymptotes. Draw through the point of contact *P* of the tangent *AB* two lines, one parallel to one asymptote and the other parallel to the other. One of these lines meets *OB* at a distance *y* from *O*, and the other meets *OA* at a distance *x* from *O*. Then, since *P* is the middle point of *AB*, *x* is one half of *OA* and *y* is one half of *OB*. The area of the parallelogram whose adjacent sides are *x* and *y* is one half the area of the triangle AOB, and therefore, by the preceding paragraph, is constant. This area is equal to $xy \cdot \sin a$, where α is the constant angle between the asymptotes. It follows that the product xy is constant, and since x and y are the oblique coördinates of the point P, the asymptotes being the axes of reference, we have

The equation of the hyperbola, referred to the asymptotes as axes, is xy = constant.

This identifies the curve with the hyperbola as defined and discussed in works on analytic geometry.

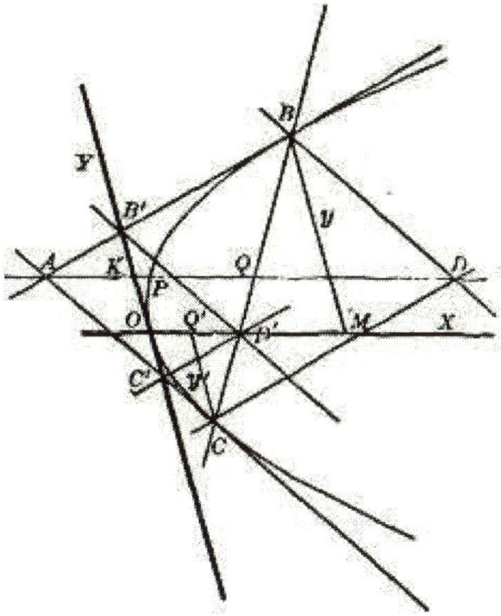

FIG. 30

119. Equation of parabola. We have defined the parabola as a conic which is tangent to the line at infinity (§ 110). Draw now two tangents to the curve (Fig. 30), meeting in A, the points of contact being B and C. These two tangents, together with the line at infinity, form a triangle circumscribed about the conic. Draw through B a parallel to AC, and through C a parallel to AB. If these meet in D, then AD is a diameter. Let AD meet the curve in P, and the chord BC in Q. P is then the middle point of AQ. Also, Q is the middle point of the chord BC, and therefore the diameter AD bisects all chords parallel to BC. In particular, AD passes through P, the point of contact of the tangent drawn parallel to BC.

Draw now another tangent, meeting AB in B' and AC in C'. Then these three, with the line at infinity, make a circumscribed quadrilateral. But, by Brianchon's theorem applied to a quadrilateral (§ 88), it appears that a parallel to AC through B', a parallel to AB through C', and the line BC meet in a point D'. Also, from the similar triangles $BB'D'$ and BAC we have, for all positions of the tangent line $B'C$,

$$B'D' : BB' = AC : AB,$$

or, since $B'D' = AC'$,

$$AC' : BB' = AC : AB = \text{constant}.$$

If another tangent meet AB in B'' and AC in C'', we have

$$AC' : BB' = AC'' : BB'',$$

and by subtraction we get

$$C'C'' : B'B'' = \text{constant};$$

whence

The segments cut off on any two tangents to a parabola by a variable tangent are proportional.

If now we take the tangent $B'C'$ as axis of ordinates, and the diameter through the point of contact O as axis of abscissas, calling the coordinates of $B(x, y)$ and of $C(x', y')$, then, from the similar triangles BMD' and we have

$$y : y' = BD' : D'C = BB' : AB'.$$

Also

$$y : y' = B'D' : C'C = AC' : C'C.$$

If now a line is drawn through A parallel to a diameter, meeting the axis of ordinates in K, we have

$$AK : OQ' = AC' : CC' = y : y',$$

and

$$OM : AK = BB' : AB' = y : y',$$

and, by multiplication,

$$OM : OQ' = y^2 : y'^2,$$

or

$$x : x' = y^2 : y'^2;$$

whence

The abscissas of two points on a parabola are to each other as the squares of the corresponding coördinates, a diameter and the tangent to the curve at the extremity of the diameter being the axes of reference.

The last equation may be written

$$y^2 = 2px,$$

where $2p$ stands for $y'^2 : x'$.

The parabola is thus identified with the curve of the same name studied in treatises on analytic geometry.

120. Equation of central conics referred to conjugate diameters. Consider now a *central conic*, that is, one which is not a parabola and the center of which is therefore at a finite distance. Draw any four tangents to it, two of which are parallel (Fig. 31). Let the parallel tangents meet one of the other tangents in A and B and the other in C and D, and let P and Q be the points of contact of the parallel tangents R and S of the others. Then AC, BD, PQ, and RS all meet in a point W (§ 88). From the figure,

$$PW : WQ = AP : QC = PD : BQ,$$

or

$$AP \cdot BQ = PD \cdot QC.$$

If now DC is a fixed tangent and AB a variable one, we have from this equation

$$AP \cdot BQ = \text{constant}.$$

This constant will be positive or negative according as PA and BQ are measured in the same or in opposite directions. Accordingly we write

$$AP \cdot BQ = \pm\, b^2.$$

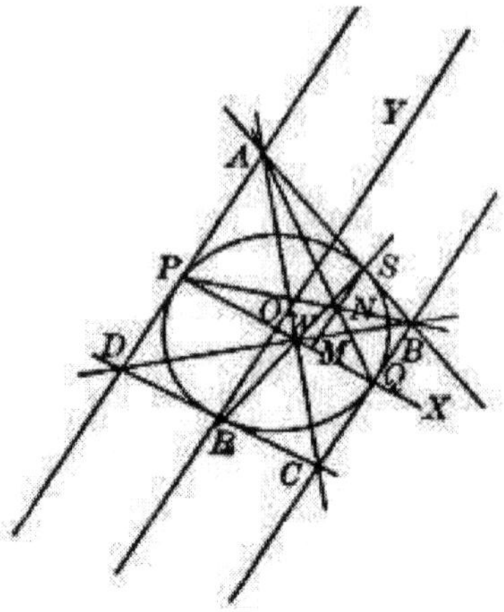

FIG. 31

Since AD and BC are parallel tangents, PQ is a diameter and the conjugate diameter is parallel to AD. The middle point of PQ is the center of the conic. We take now for the axis of abscissas the diameter PQ, and the conjugate diameter for the axis of ordinates. Join A to Q and B to P and draw a line through S parallel to the axis of ordinates. These three lines all meet in a point N, because AP, BQ, and AB form a triangle circumscribed to the conic. Let NS meet PQ in M. Then, from the properties of the circumscribed triangle (§ 89), M, N, S, and the point at infinity on NS are four harmonic points, and therefore N is the middle point of MS. If the coördinates of S are (x, y), so that OM is x and MS is y, then $MN = y/2$. Now from the similar triangles PMN and PQB we have

$$BQ : PQ = NM : PM,$$

and from the similar triangles PQA and MQN,

$$AP : PQ = MN : MQ,$$

whence, multiplying, we have

$$\pm b^2/4\, a^2 = y^2/4\, (a + x)(a - x),$$

where

$$a = \frac{PQ}{2},$$

or, simplifying,

$$x^2/a^2 + y^2/\pm b^2 = 1,$$

which is the equation of an ellipse when b^2 has a positive sign, and of a hyperbola when b^2 has a negative sign. We have thus identified point-rows of the second order with the curves given by equations of the second degree.

PROBLEMS

1. Draw a chord of a given conic which shall be bisected by a given point *P*.

2. Show that all chords of a given conic that are bisected by a given chord are tangent to a parabola.

3. Construct a parabola, given two tangents with their points of contact.

4. Construct a parabola, given three points and the direction of the diameters.

5. A line u' is drawn through the pole *U* of a line *u* and at right angles to *u*. The line *u* revolves about a point *P*. Show that the line u' is tangent to a parabola. (The lines *u* and u' are called normal conjugates.)

6. Given a circle and its center *O*, to draw a line through a given point *P* parallel to a given line *q*. Prove the following construction: Let *p* be the polar of *P*, *Q* the pole of *q*, and *A* the intersection of *p* with *OQ*. The polar of *A* is the desired line.

CHAPTER VIII
INVOLUTION

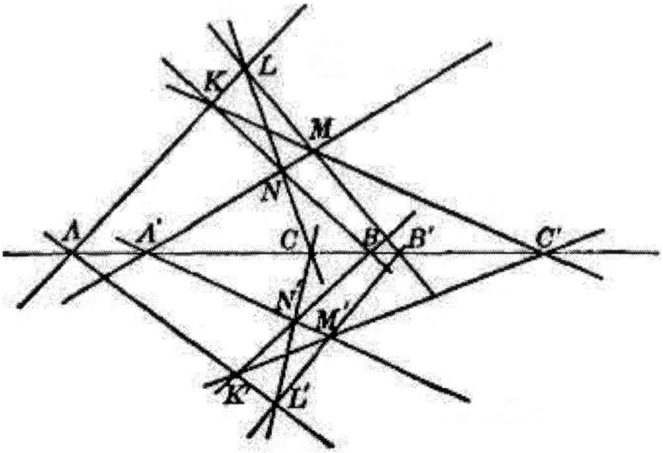

FIG. 32

121. Fundamental theorem. The important theorem concerning two complete quadrangles (§ 26), upon which the theory of four harmonic points was based, can easily be extended to the case where the four lines *KL*, *K'L'*, *MN*, *M'N'* do not all meet in the same point *A*, and the more general theorem that results may also be made the basis of a theory no less important, which has to do with six points on a line. The theorem is as follows:

Given two complete quadrangles, K, L, M, N and K', L', M', N', so related that KL and K'L' meet in A, MN and M'N' in A', KN and K'N' in B, LM and L'M' in B', LN and L'N' in C, and KM and K'M' in C', then, if A, A', B, B', and C are in a straight line, the point C' also lies on that straight line.

The theorem follows from Desargues's theorem (Fig. 32). It is seen that *KK'*, *LL'*, *MM'*, *NN'* all meet in a point, and thus, from the same theorem, applied to the triangles *KLM* and *K'L'M'*, the point *C'* is on the same line with *A* and *B'*. As in the simpler case, it is seen that there is an indefinite number of quadrangles which may be drawn, two sides of which go through *A* and *A'*, two through *B* and *B'*, and one through *C*. The sixth side must then go through *C'*. Therefore,

122. *Two pairs of points, A, A′ and B, B′, being given, then the point C′ corresponding to any given point C is uniquely determined.*

The construction of this sixth point is easily accomplished. Draw through *A* and *A′* any two lines, and cut across them by any line through *C* in the points *L* and *N*. Join *N* to *B* and *L* to *B′*, thus determining the points *K* and *M* on the two lines through *A* and *A′*, The line *KM* determines the desired point *C′*. Manifestly, starting from *C′*, we come in this way always to the same point *C*. The particular quadrangle employed is of no consequence. Moreover, since one pair of opposite sides in a complete quadrangle is not distinguishable in any way from any other, the same set of six points will be obtained by starting from the pairs *AA′* and *CC′*, or from the pairs *BB′* and *CC′*.

123. Definition of involution of points on a line.

Three pairs of points on a line are said to be in involution if through each pair may be drawn a pair of opposite sides of a complete quadrangle. If two pairs are fixed and one of the third pair describes the line, then the other also describes the line, and the points of the line are said to be paired in the involution determined by the two fixed pairs.

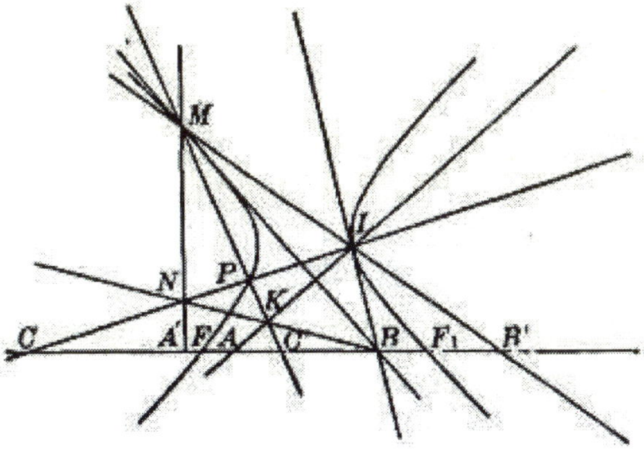

FIG. 33

124. Double-points in an involution. The points C and C' describe projective point-rows, as may be seen by fixing the points L and M. The self-corresponding points, of which there are two or none, are called the *double-points* in the involution. It is not difficult to see that the double-points in the involution are harmonic conjugates with respect to corresponding points in the involution. For, fixing as before the points L and M, let the intersection of the lines CL and $C'M$ be P (Fig. 33). The locus of P is a conic which goes through the double-points, because the point-rows C and C' are projective, and therefore so are the pencils LC and MC' which generate the locus of P. Also, when C and C' fall together, the point P coincides with them. Further, the tangents at L and M to this conic described by P are the lines LB and MB. For in the pencil at L the ray LM common to the two pencils which generate the conic is the ray LB' and corresponds to the ray MB of M, which is therefore the tangent line to the conic at M. Similarly for the tangent LB at L. LM is therefore the polar of B with respect to this conic, and B and B' are therefore harmonic conjugates with respect to the double-points. The same discussion applies to any other pair of corresponding points in the involution.

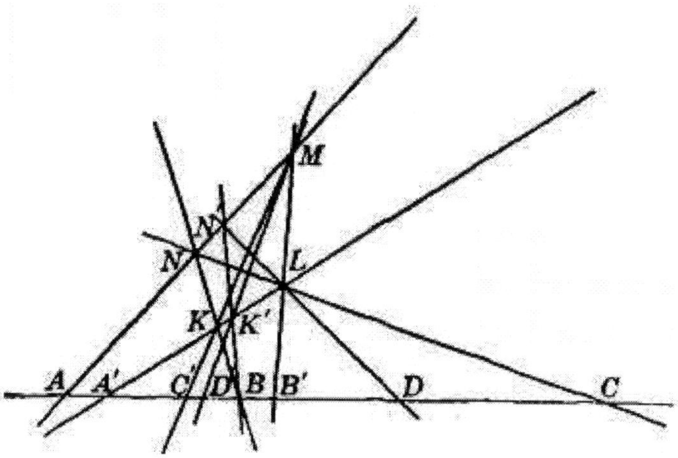

FIG. 34

125. Desargues's theorem concerning conics through four points. Let DD' be any pair of points in the involution determined as above, and consider the conic passing through the five points K, L, M, N, D. We shall use Pascal's theorem to show that this conic also passes through D'. The point D' is determined as follows: Fix L and M as before (Fig. 34) and join D to L, giving on MN the point N'. Join N' to B, giving on LK the point K'. Then MK' determines the point D' on the line AA', given by the complete quadrangle K', L, M, N'. Consider the following six points, numbering them in order: $D = 1$, $D' = 2$, $M = 3$, $N = 4$, $K = 5$, and $L = 6$. We have the following intersections: $B = (12\text{-}45)$, $K' = (23\text{-}56)$, $N' = (34\text{-}61)$; and since by construction B, N, and K' are on a straight line, it follows from the converse of Pascal's theorem, which is easily established, that the six points are on a conic. We have, then, the beautiful theorem due to Desargues:

The system of conics through four points meets any line in the plane in pairs of points in involution.

126. It appears also that the six points in involution determined by the quadrangle through the four fixed points belong also to the same involution with the points cut out by the system of conics, as indeed we might infer from the fact that the three pairs of opposite sides of the quadrangle may be considered as degenerate conics of the system.

127. Conics through four points touching a given line. It is further evident that the involution determined on a line by the system of conics will have a double-point where a conic of the system is tangent to the line. We may therefore infer the theorem

Through four fixed points in the plane two conics or none may be drawn tangent to any given line.

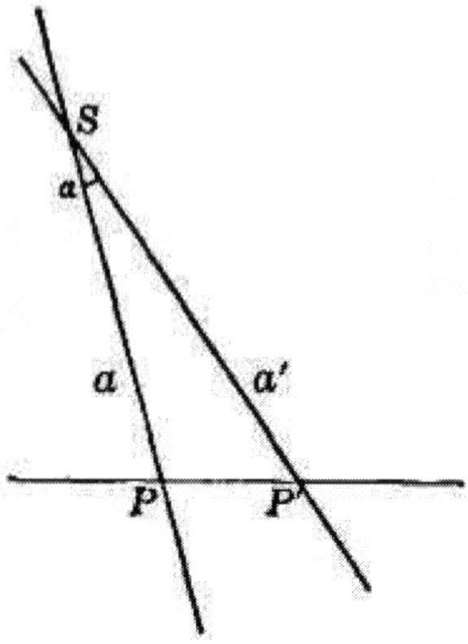

FIG. 35

128. Double correspondence. We have seen that corresponding points in an involution form two projective point-rows superposed on the same straight line. Two projective point-rows superposed on the same straight line are, however, not necessarily in involution, as a simple example will show. Take two lines, *a* and *a'*, which both revolve about a fixed point *S* and which always make the same angle with each other (Fig. 35). These lines cut out on any line in the plane which does not pass through *S* two projective point-rows, which are not, however, in involution unless the angle between the lines is a right angles. For a point *P* may correspond to a point *P'*, which in turn will correspond to some other point than *P*. The peculiarity of point-rows in involution is that any point will correspond to the same point, in whichever point-row it is considered as belonging. In this case, if a point *P* corresponds to a point *P'*, then the point *P'* corresponds back again to the point *P*. The points *P* and *P'* are then said to *correspond doubly*. This notion is worthy of further study.

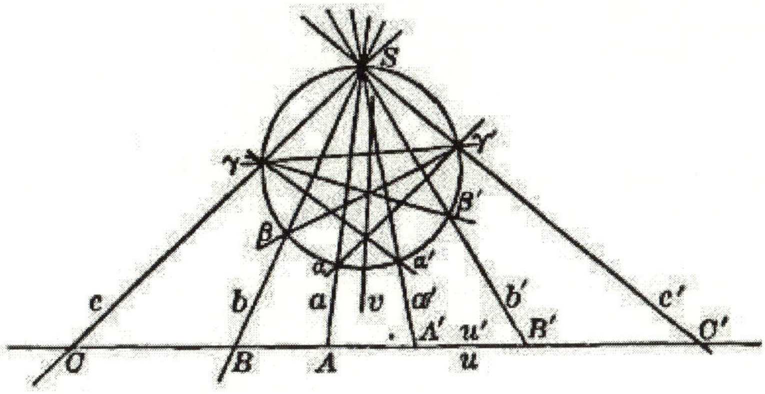

FIG. 36

129. Steiner's construction. It will be observed that the solution of the fundamental problem given in § 83, *Given three pairs of points of two protective point-rows, to construct other pairs*, cannot be carried out if the two point-rows lie on the same straight line. Of course the method may be easily altered to cover that case also, but it is worth while to give another solution of the problem, due to Steiner, which will also give further information regarding the theory of involution, and which may, indeed, be used as a foundation for that theory. Let the two point-rows *A*, *B*, *C*, *D*, ... and *A'*, *B'*, *C'*, *D'*, ... be superposed on the line *u*. Project them both to a point *S* and pass any conic *ϰ* through *S*. We thus obtain two projective pencils, *a*, *b*, *c*, *d*, ... and *a'*, *b'*, *c'*, *d'*, ... at *S*, which meet the conic in the points *α*, *β*, *γ*, *δ*, ... and *α'*, *β'*, *γ'*, *δ'*, ... (Fig. 36). Take now *γ* as the center of a pencil projecting the points *α'*, *β'*, *δ'*, ..., and take *γ'* as the center of a pencil projecting the points *α*, *β*, *δ*, ... These two pencils are projective to each other, and since they have a self-correspondin ray in common, they are in perspective position and corresponding rays meet on the line joining *(γα'*, *γ'α)* to *(γβ'*, *γ'β)*. The correspondence between points in the two point-rows on *u* is now easily traced.

130. Application of Steiner's construction to double correspondence. Steiner's construction throws into our hands an important theorem concerning double correspondence: *If two projective point-rows, superposed on the same line, have one pair of points which correspond to each other doubly, then all pairs correspond to each other doubly, and the line is paired in involution.* To make this appear, let us call the point A on u by two names, A and P', according as it is thought of as belonging to the one or to the other of the two point-rows. If this point is one of a pair which correspond to each other doubly, then the points A' and P must coincide (Fig. 37). Take now any point C, which we will also call R'. We must show that the corresponding point C' must also coincide with the point B. Join all the points to S, as before, and it appears that the points α and π' coincide, as also do the points $a'\pi$ and $\gamma\varrho'$. By the above construction the line $\gamma'\varrho$ must meet $\gamma\varrho'$ on the line joining $(\gamma a', \gamma' a)$ with $(\gamma\pi', \gamma'\pi)$. But these four points form a quadrangle inscribed in the conic, and we know by § 95 that the tangents at the opposite vertices γ and γ' meet on the line v. The line $\gamma'\varrho$ is thus a tangent to the conic, and C' and R are the same point. That two projective point-rows superposed on the same line are also in involution when one pair, and therefore all pairs, correspond doubly may be shown by taking S at one vertex of a complete quadrangle which has two pairs of opposite sides going through two pairs of points. The details we leave to the student.

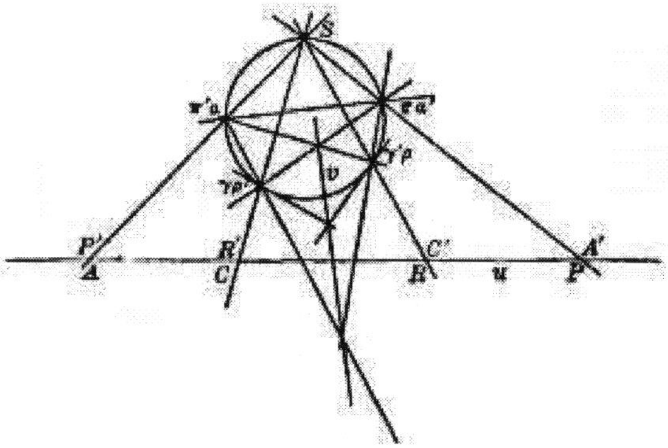

FIG. 37

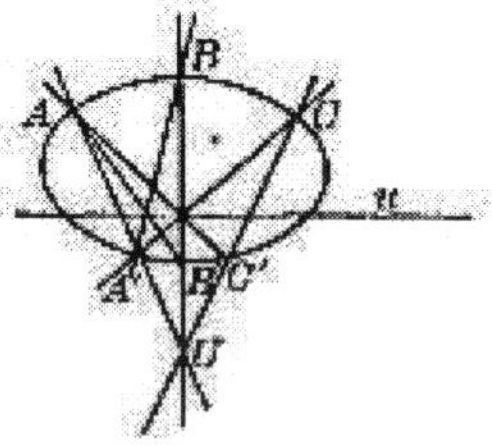

FIG. 38

131. Involution of points on a point-row of the second order. It is important to note also, in Steiner's construction, that we have obtained two point-rows of the second order superposed on the same conic, and have paired the points of one with the points of the other in such a way that the correspondence is double. We may then extend the notion of involution to point-rows of the second order and say that *the points of a conic are paired in involution when they are corresponding points of two projective point-rows superposed on the conic, and when they correspond to each other doubly.* With this definition we may prove the theorem: *The lines joining corresponding points of a point-row of the second order in involution all pass through a fixed point U, and the line joining any two points A, B meets the line joining the two corresponding points A', B' in the points of a line u, which is the polar of U with respect to the conic.* For take A and A' as the centers of two pencils, the first perspective to the point-row A', B', C' and the second perspective to the point-row A, B, C. Then, since the common ray of the two pencils corresponds to itself, they are in perspective position, and their axis of perspectivity u (Fig. 38) is the line which joins the point $(AB', A'B)$ to the point $(AC', A'C)$. It is then immediately clear, from the theory of poles and polars, that BB' and CC' pass through the pole U of the line u.

132. Involution of rays. The whole theory thus far developed may be dualized, and a theory of lines in involution may be built up, starting with the complete quadrilateral. Thus,

The three pairs of rays which may be drawn from a point through the three pairs of opposite vertices of a complete quadrilateral are said to be in involution. If the pairs aa' and bb' are fixed, and the line c describes a pencil, the corresponding line c' also describes a pencil, and the rays of the pencil are said to be paired in the involution determined by aa' and bb'.

133. Double rays. The self-corresponding rays, of which there are two or none, are called *double rays* of the involution. Corresponding rays of the involution are harmonic conjugates with respect to the double rays. To the theorem of Desargues (§ 125) which has to do with the system of conics through four points we have the dual:

The tangents from a fixed point to a system of conics tangent to four fixed lines form a pencil of rays in involution.

134. If a conic of the system should go through the fixed point, it is clear that the two tangents would coincide and indicate a double ray of the involution. The theorem, therefore, follows:

Two conics or none may be drawn through a fixed point to be tangent to four fixed lines.

135. Double correspondence. It further appears that two projective pencils of rays which have the same center are in involution if two pairs of rays correspond to each other doubly. From this it is clear that we might have deemed six rays in involution as six rays which pass through a point and also through six points in involution. While this would have been entirely in accord with the treatment which was given the corresponding problem in the theory of harmonic points and lines, it is more satisfactory, from an aesthetic point of view, to build the theory of lines in involution on its own base. The student can show, by methods entirely analogous to those used in the second chapter, that involution is a projective property; that is, six rays in involution are cut by any transversal in six points in involution.

136. Pencils of rays of the second order in involution. We may also extend the notion of involution to pencils of rays of the second order. Thus, *the tangents to a conic are in involution when they are corresponding rays of two protective pencils of the second order superposed upon the same conic, and when they correspond to each other doubly.* We have then the theorem:

137. *The intersections of corresponding rays of a pencil of the second order in involution are all on a straight line u, and the intersection of any two tangents ab, when joined to the intersection of the corresponding tangents a'b', gives a line which passes through a fixed point U, the pole of the line u with respect to the conic.*

138. Involution of rays determined by a conic. We have seen in the theory of poles and polars (§ 103) that if a point *P* moves along a line *m*, then the polar of P revolves about a point. This pencil cuts out on m another point-row P', projective also to P. Since the polar of P passes through P', the polar of P' also passes through P, so that the correspondence between P and P' is double. The two point-rows are therefore in involution, and the double points, if any exist, are the points where the line m meets the conic. A similar involution of rays may be found at any point in the plane, corresponding rays passing each through the pole of the other. We have called such points and rays *conjugate* with respect to the conic (§ 100). We may then state the following important theorem:

139. *A conic determines on every line in its plane an involution of points, corresponding points in the involution being conjugate with respect to the conic. The double points, if any exist, are the points where the line meets the conic.*

140. The dual theorem reads: *A conic determines at every point in the plane an involution of rays, corresponding rays being conjugate with respect to the conic. The double rays, if any exist, are the tangents from the point to the conic.*

PROBLEMS

1. Two lines are drawn through a point on a conic so as always to make right angles with each other. Show that the lines joining the points where they meet the conic again all pass through a fixed point.

2. Two lines are drawn through a fixed point on a conic so as always to make equal angles with the tangent at that point. Show that the lines joining the two points where the lines meet the conic again all pass through a fixed point.

3. Four lines divide the plane into a certain number of regions. Determine for each region whether two conics or none may be drawn to pass through points of it and also to be tangent to the four lines.

4. If a variable quadrangle move in such a way as always to remain inscribed in a fixed conic, while three of its sides turn each around one of three fixed collinear points, then the fourth will also turn around a fourth fixed point collinear with the other three.

5. State and prove the dual of problem 4.

6. Extend problem 4 as follows: If a variable polygon of an even number of sides move in such a way as always to remain inscribed in a fixed conic, while all its sides but one pass through as many fixed collinear points, then the last side will also pass through a fixed point collinear with the others.

7. If a triangle *QRS* be inscribed in a conic, and if a transversal *s* meet two of its sides in *A* and *A′*, the third side and the tangent at the opposite vertex in *B* and *B′*, and the conic itself in *C* and *C′*, then *AA′*, *BB′*, *CC′* are three pairs of points in an involution.

8. Use the last exercise to solve the problem: Given five points, *Q*, *R*, *S*, *C*, *C′*, on a conic, to draw the tangent at any one of them.

9. State and prove the dual of problem 7 and use it to prove the dual of problem 8.

10. If a transversal cut two tangents to a conic in *B* and *B′*, their chord of contact in *A*, and the conic itself in *P* and *P′*, then the point *A* is a double point of the involution determined by *BB′* and *PP′*.

11. State and prove the dual of problem 10.

12. If a variable conic pass through two given points, *P* and *P′*, and if it be tangent to two given lines, the chord of contact of these two tangents will always pass through a fixed point on *PP′*.

13. Use the last theorem to solve the problem: Given four points, *P*, *P′*, *Q*, *S*, on a conic, and the tangent at one of them, *Q*, to draw the tangent at any one of the other points, *S*.

14. Apply the theorem of problem 9 to the case of a hyperbola where the two tangents are the asymptotes. Show in this way that if a hyperbola and its asymptotes be cut by a transversal, the segments intercepted by the curve and by the asymptotes respectively have the same middle point.

15. In a triangle circumscribed about a conic, any side is divided harmonically by its point of contact and the point where it meets the chord joining the points of contact of the other two sides.

CHAPTER IX
METRICAL PROPERTIES OF INVOLUTIONS

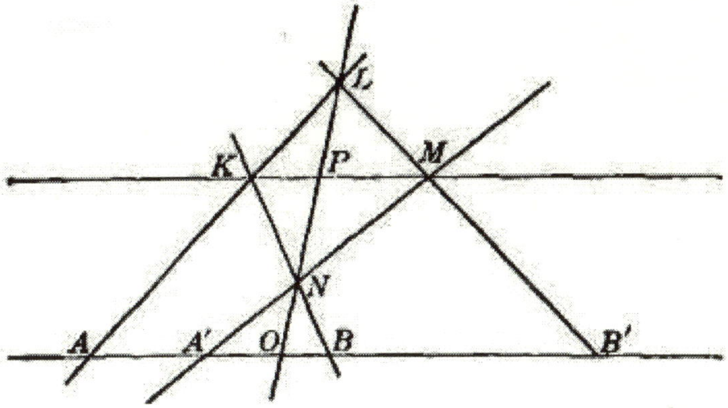

FIG. 39

141. Introduction of infinite point; center of involution. We connect the projective theory of involution with the metrical, as usual, by the introduction of the elements at infinity. In an involution of points on a line the point which corresponds to the infinitely distant point is called the *center* of the involution. Since corresponding points in the involution have been shown to be harmonic conjugates with respect to the double points, the center is midway between the double points when they exist. To construct the center (Fig. 39) we draw as usual through A and A' any two rays and cut them by a line parallel to AA' in the points K and M. Join these points to B and B', thus determining on AK and AN the points L and N. LN meets AA' in the center O of the involution.

142. Fundamental metrical theorem. From the figure we see that the triangles OLB' and PLM are similar, P being the intersection of KM and LN. Also the triangles KPN and BON are similar. We thus have

$$OB : PK = ON : PN$$

and

$$OB' : PM = OL : PL;$$

whence

$$OB \cdot OB' : PK \cdot PM = ON \cdot OL : PN \cdot PL.$$

In the same way, from the similar triangles OAL and PKL, and also $OA'N$ and PMN, we obtain

$$OA \cdot OA' : PK \cdot PM = ON \cdot OL : PN \cdot PL,$$

and this, with the preceding, gives at once the fundamental theorem, which is sometimes taken also as the definition of involution:

$$OA \cdot OA' = OB \cdot OB' = \text{constant},$$

or, in words,

The product of the distances from the center to two corresponding points in an involution of points is constant.

143. Existence of double points. Clearly, according as the constant is positive or negative the involution will or will not have double points. The constant is the square root of the distance from the center to the double points. If A and A' lie both on the same side of the center, the product $OA \cdot OA'$ is positive; and if they lie on opposite sides, it is negative. Take the case where they both lie on the same side of the center, and take also the pair of corresponding points BB'. Then, since $OA \cdot OA' = OB \cdot OB'$, it cannot happen that B and B' are separated from each other by A and A'. This is evident enough if the points are on opposite sides of the center. If the pairs are on the same side of the center, and B lies between A and A', so that OB is greater, say, than OA, but less than OA', then, by the equation $OA \cdot OA' = OB \cdot OB'$, we must have OB' also less than OA' and greater than OA. A similar discussion may be made for the case where A and A' lie on opposite sides of O. The results may be stated as follows, without any reference to the center:

Given two pairs of points in an involution of points, if the points of one pair are separated from each other by the points of the other pair, then the involution has no double points. If the points of one pair are not separated from each other by the points of the other pair, then the involution has two double points.

144. An entirely similar criterion decides whether an involution of rays has or has not double rays, or whether an involution of planes has or has not double planes.

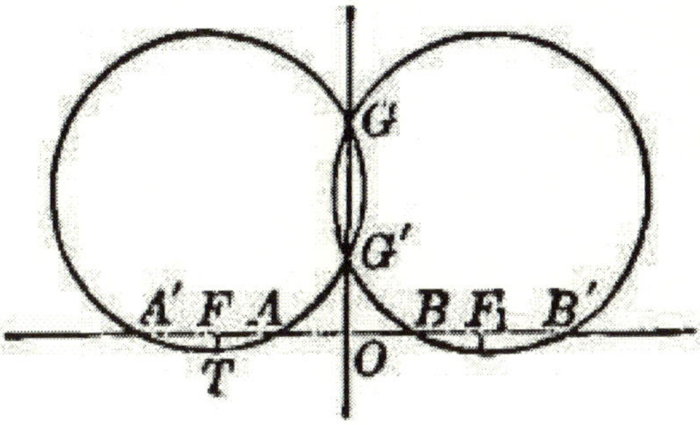

FIG. 40

145. Construction of an involution by means of circles. The equation just derived, $OA \cdot OA' = OB \cdot OB'$, indicates another simple way in which points of an involution of points may be constructed. Through A and A' draw any circle, and draw also any circle through B and B' to cut the first in the two points G and G' (Fig. 40). Then any circle through G and G' will meet the line in pairs of points in the involution determined by AA' and BB'. For if such a circle meets the line in the points CC', then, by the theorem in the geometry of the circle which says that *if any chord is drawn through a fixed point within a circle, the product of its segments is constant in whatever direction the chord is drawn, and if a secant line be drawn from a fixed point without a circle, the product of the secant and its external segment is constant in whatever direction the secant line is drawn*, we have $OC \cdot OC' = OG \cdot OG' =$ constant. So that for all such points $OA \cdot OA' = OB \cdot OB' = OC \cdot OC'$. Further, the line GG' meets AA' in the center of the involution. To find the double points, if they exist, we draw a tangent from O to any of the circles through GG'. Let T be the point of contact. Then lay off on the line OA a line OF equal to OT. Then, since by the above theorem of elementary geometry $OA \cdot OA' = OT^2 = OF^2$, we have one double point F. The other is at an equal distance on the other side of O. This simple and effective method of constructing an involution of points is often taken as the basis for the theory of involution. In projective geometry, however, the circle, which is not a figure that remains unaltered by projection, and is essentially a metrical notion, ought not to be used to build up the purely projective part of the theory.

146. It ought to be mentioned that the theory of analytic geometry indicates that the circle is a special conic section that happens to pass through two particular imaginary points on the line at infinity, called the *circular points* and usually denoted by I and J. The above method of obtaining a point-row in involution is, then, nothing but a special case of the general theorem of the last chapter (§ 125), which asserted that a system of conics through four points will cut any line in the plane in a point-row in involution.

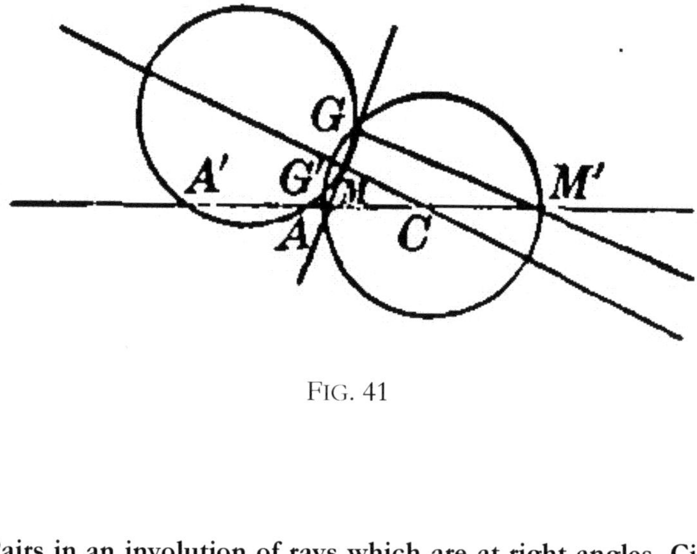

FIG. 41

147. Pairs in an involution of rays which are at right angles. Circular involution. In an involution of rays there is no one ray which may be distinguished from all the others as the point at infinity is distinguished from all other points on a line. There is one pair of rays, however, which does differ from all the others in that for this particular pair the angle is a right angle. This is most easily shown by using the construction that employs circles, as indicated above. The centers of all the circles through G and G' lie on the perpendicular bisector of the line GG'. Let this line meet the line AA' in the point C (Fig. 41), and draw the circle with center C which goes through G and G'. This circle cuts out two points M and M' in the involution. The rays GM and GM' are clearly at right angles, being inscribed in a semicircle. If, therefore, the involution of points is projected to G, we have found two corresponding rays which are at right angles to each other. Given now any involution of rays with center G, we may cut across it by a straight line and proceed to find the two points M and M'. Clearly there will be only one such pair unless the perpendicular bisector of GG' coincides with the line AA'. In this case every ray is at right angles to its corresponding ray, and the involution is called *circular*.

148. Axes of conics. At the close of the last chapter (§ 140) we gave the theorem: *A conic determines at every point in its plane an involution of rays, corresponding rays being conjugate with respect to the conic. The double rays, if any exist, are the tangents from the point to the conic.* In particular, taking the point as the center of the conic, we find that conjugate diameters form a system of rays in involution, of which the asymptotes, if there are any, are the double rays. Also, conjugate diameters are harmonic conjugates with respect to the asymptotes. By the theorem of the last paragraph, there are two conjugate diameters which are at right angles to each other. These are called axes. In the case of the parabola, where the center is at infinity, and on the curve, there are, properly speaking, no conjugate diameters. While the line at infinity might be considered as conjugate to all the other diameters, it is not possible to assign to it any particular direction, and so it cannot be used for the purpose of defining an axis of a parabola. There is one diameter, however, which is at right angles to its conjugate system of chords, and this one is called the *axis* of the parabola. The circle also furnishes an exception in that every diameter is an axis. The involution in this case is circular, every ray being at right angles to its conjugate ray at the center.

149. Points at which the involution determined by a conic is circular. It is an important problem to discover whether for any conic other than the circle it is possible to find any point in the plane where the involution determined as above by the conic is circular. We shall proceed to the curious problem of proving the existence of such points and of determining their number and situation. We shall then develop the important properties of such points.

150. It is clear, in the first place, that such a point cannot be on the outside of the conic, else the involution would have double rays and such rays would have to be at right angles to themselves. In the second place, if two such points exist, the line joining them must be a diameter and, indeed, an axis. For if F and F' were two such points, then, since the conjugate ray at F to the line FF' must be at right angles to it, and also since the conjugate ray at F' to the line FF' must be at right angles to it, the pole of FF' must be at infinity in a direction at right angles to FF'. The line FF' is then a diameter, and since it is at right angles to its conjugate diameter, it must be an axis. From this it follows also that the points we are seeking must all lie on one of the two axes, else we should have a diameter which does not go through the intersection of all axes—the center of the conic. At least one axis, therefore, must be free from any such points.

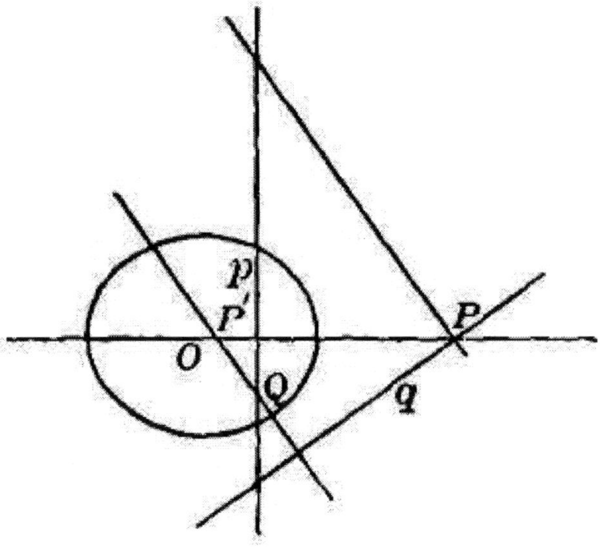

FIG. 42

151. Let now P be a point on one of the axes (Fig. 42), and draw any ray through it, such as q. As q revolves about P, its pole Q moves along a line at right angles to the axis on which P lies, describing a point-row p projective to the pencil of rays q. The point at infinity in a direction at right angles to q also describes a point-row projective to q. The line joining corresponding points of these two point-rows is always a conjugate line to q and at right angles to q, or, as we may call it, a *conjugate normal* to q. These conjugate normals to q, joining as they do corresponding points in two projective point-rows, form a pencil of rays of the second order. But since the point at infinity on the point-row Q corresponds to the point at infinity in a direction at right angles to q, these point-rows are in perspective position and the normal conjugates of all the lines through P meet in a point. This point lies on the same axis with P, as is seen by taking q at right angles to the axis on which P lies. The center of this pencil may be called P', and thus we have paired the point P with the point P'. By moving the point P along the axis, and by keeping the ray q parallel to a fixed direction, we may see that the point-row P and the point-row P' are projective. Also the correspondence is double, and by starting from the point P' we arrive at the point P. Therefore the point-rows P and P' are in involution, and if only the involution has double points, we shall have found in them the points we are seeking. For it is clear that the rays through P and the corresponding rays through P' are conjugate normals; and if P and P' coincide, we shall have a point where all rays are at right angles to their conjugates. We shall now show that the involution thus obtained on one of the two axes must have double points.

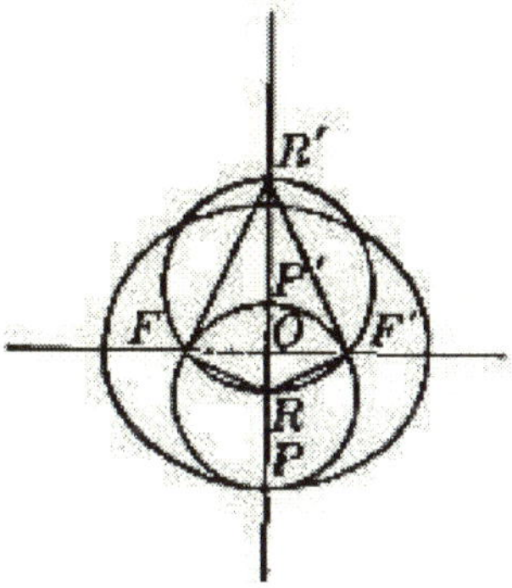

FIG. 43

152. Discovery of the foci of the conic. We know that on one axis no such points as we are seeking can lie (§ 150). The involution of points PP' on this axis can therefore have no double points. Nevertheless, let PP' and RR' be two pairs of corresponding points on this axis (Fig. 43). Then we know that P and P' are separated from each other by R and R' (§ 143). Draw a circle on PP' as a diameter, and one on RR' as a diameter. These must intersect in two points, F and F', and since the center of the conic is the center of the involution PP', RR', as is easily seen, it follows that F and F' are on the other axis of the conic. Moreover, FR and FR' are conjugate normal rays, since RFR' is inscribed in a semicircle, and the two rays go one through R and the other through R'. The involution of points PP', RR' therefore projects to the two points F and F' in two pencils of rays in involution which have for corresponding rays conjugate normals to the conic. We may, then, say:

There are two and only two points of the plane where the involution determined by the conic is circular. These two points lie on one of the axes, at equal distances from the center, on the inside of the conic. These points are called the foci of the conic.

153. The circle and the parabola. The above discussion applies only to the central conics, apart from the circle. In the circle the two foci fall together at the center. In the case of the parabola, that part of the investigation which proves the existence of two foci on one of the axes will not hold, as we have but one axis. It is seen, however, that as P moves to infinity, carrying the line q with it, q becomes the line at infinity, which for the parabola is a tangent line. Its pole Q is thus at infinity and also the point P', so that P and P' fall together at infinity, and therefore one focus of the parabola is at infinity. There must therefore be another, so that

A parabola has one and only one focus in the finite part of the plane.

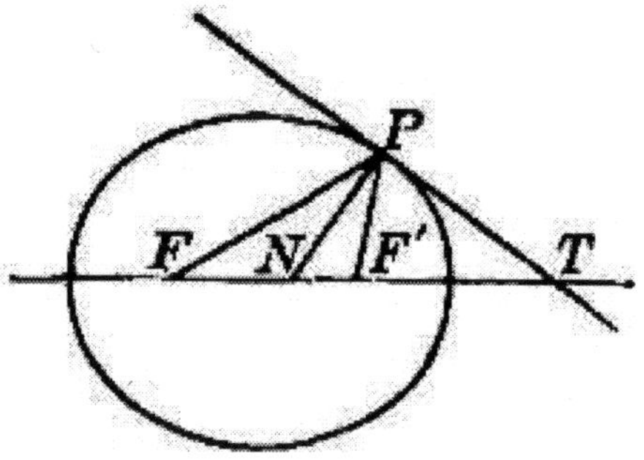

FIG. 44

154. Focal properties of conics. We proceed to develop some theorems which will exhibit the importance of these points in the theory of the conic section. Draw a tangent to the conic, and also the normal at the point of contact *P*. These two lines are clearly conjugate normals. The two points *T* and *N*, therefore, where they meet the axis which contains the foci, are corresponding points in the involution considered above, and are therefore harmonic conjugates with respect to the foci (Fig. 44); and if we join them to the point *P*, we shall obtain four harmonic lines. But two of them are at right angles to each other, and so the others make equal angles with them (Problem 4, Chapter II). Therefore

The lines joining a point on the conic to the foci make equal angles with the tangent.

It follows that rays from a source of light at one focus are reflected by an ellipse to the other.

155. In the case of the parabola, where one of the foci must be considered to be at infinity in the direction of the diameter, we have

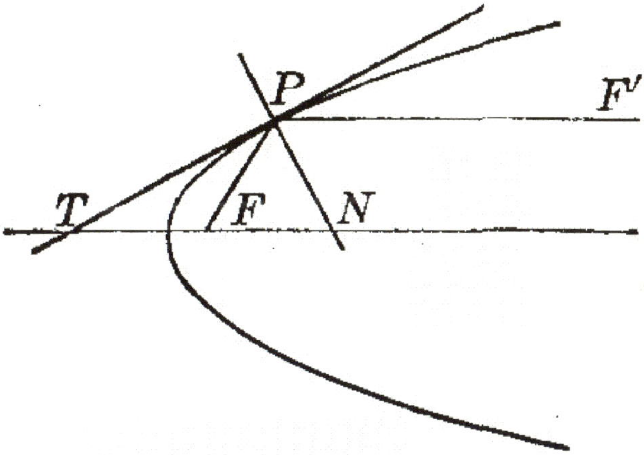

FIG. 45

A diameter makes the same angle with the tangent at its extremity as that tangent does with the line from its point of contact to the focus (Fig. 45).

156. This last theorem is the basis for the construction of the parabolic reflector. A ray of light from the focus is reflected from such a reflector in a direction parallel to the axis of the reflector.

157. Directrix. Principal axis. Vertex. The polar of the focus with respect to the conic is called the *directrix*. The axis which contains the foci is called the *principal axis*, and the intersection of the axis with the curve is called the *vertex* of the curve. The directrix is at right angles to the principal axis. In a parabola the vertex is equally distant from the focus and the directrix, these three points and the point at infinity on the axis being four harmonic points. In the ellipse the vertex is nearer to the focus than it is to the directrix, for the same reason, and in the hyperbola it is farther from the focus than it is from the directrix.

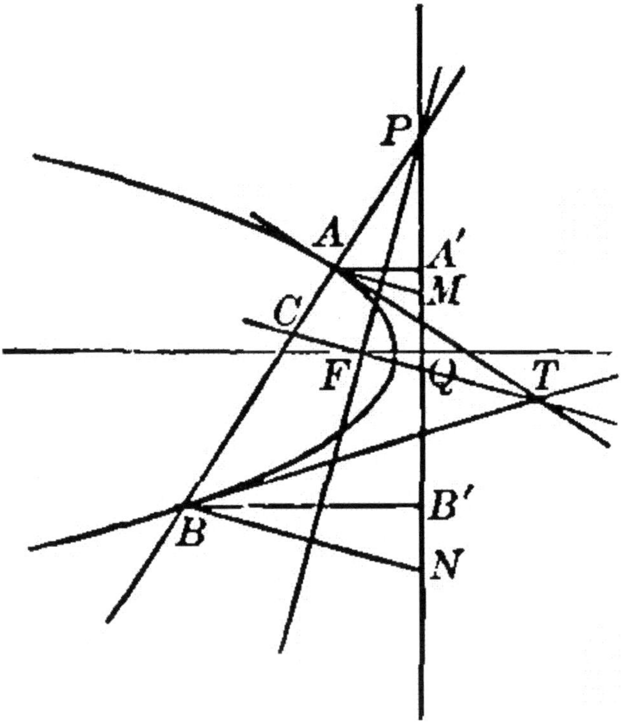

FIG. 46

158. Another definition of a conic. Let P be any point on the directrix through which a line is drawn meeting the conic in the points A and B (Fig. 46). Let the tangents at A and B meet in T, and call the focus F. Then TF and PF are conjugate lines, and as they pass through a focus they must be at right angles to each other. Let TF meet AB in C. Then P, A, C, B are four harmonic points. Project these four points parallel to TF upon the directrix, and we then get the four harmonic points P, M, Q, N. Since, now, TFP is a right angle, the angles MFQ and NFQ are equal, as well as the angles AFC and BFC. Therefore the triangles MAF and NFB are similar, and $FA : FM = FB : BN$. Dropping perpendiculars AA and BB' upon the directrix, this becomes $FA : AA' = FB : BB'$. We have thus the property often taken as the definition of a conic:

The ratio of the distances from a point on the conic to the focus and the directrix is constant.

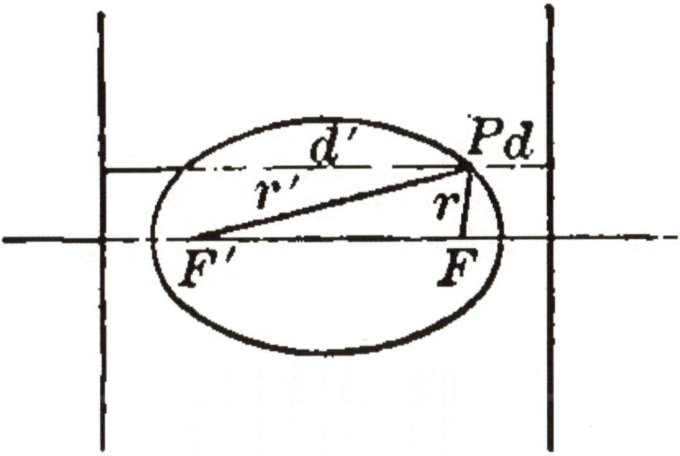

FIG. 47

159. Eccentricity. By taking the point at the vertex of the conic, we note that this ratio is less than unity for the ellipse, greater than unity for the hyperbola, and equal to unity for the parabola. This ratio is called the *eccentricity*.

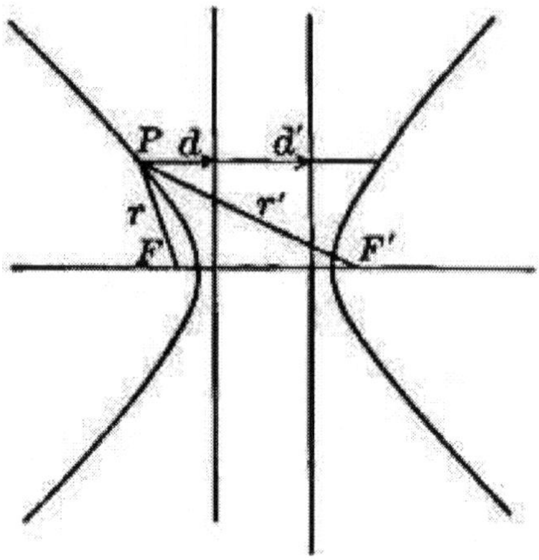

FIG. 48

160. Sum or difference of focal distances. The ellipse and the hyperbola have two foci and two directrices. The eccentricity, of course, is the same for one focus as for the other, since the curve is symmetrical with respect to both. If the distances from a point on a conic to the two foci are r and r', and the distances from the same point to the corresponding directrices are d and d' (Fig. 47), we have $r : d = r' : d'$; $(r \pm r') : (d \pm d')$. In the ellipse $(d + d')$ is constant, being the distance between the directrices. In the hyperbola this distance is $(d - d')$. It follows (Fig. 48) that

In the ellipse the sum of the focal distances of any point on the curve is constant, and in the hyperbola the difference between the focal distances is constant.

PROBLEMS

1. Construct the axis of a parabola, given four tangents.

2. Given two conjugate lines at right angles to each other, and let them meet the axis which has no foci on it in the points A and B. The circle on AB as diameter will pass through the foci of the conic.

3. Given the axes of a conic in position, and also a tangent with its point of contact, to construct the foci and determine the length of the axes.

4. Given the tangent at the vertex of a parabola, and two other tangents, to find the focus.

5. The locus of the center of a circle touching two given circles is a conic with the centers of the given circles for its foci.

6. Given the axis of a parabola and a tangent, with its point of contact, to find the focus.

7. The locus of the center of a circle which touches a given line and a given circle consists of two parabolas.

8. Let F and F' be the foci of an ellipse, and P any point on it. Produce PF to G, making PG equal to PF'. Find the locus of G.

9. If the points G of a circle be folded over upon a point F, the creases will all be tangent to a conic. If F is within the circle, the conic will be an ellipse; if F is without the circle, the conic will be a hyperbola.

10. If the points G in the last example be taken on a straight line, the locus is a parabola.

11. Find the foci and the length of the principal axis of the conics in problems 9 and 10.

12. In problem 10 a correspondence is set up between straight lines and parabolas. As there is a fourfold infinity of parabolas in the plane, and only a twofold infinity of straight lines, there must be some restriction on the parabolas obtained by this method. Find and explain this restriction.

13. State and explain the similar problem for problem 9.

14. The last four problems are a study of the consequences of the following transformation: A point O is fixed in the plane. Then to any point P is made to correspond the line p at right angles to OP and bisecting it. In this correspondence, what happens to p when P moves along a straight line? What corresponds to the theorem that two lines have only one point in common? What to the theorem that the angle sum of a triangle is two right angles? Etc.

CHAPTER X
ON THE HISTORY OF SYNTHETIC PROJECTIVE GEOMETRY

161. Ancient results. The theory of synthetic projective geometry as we have built it up in this course is less than a century old. This is not to say that many of the theorems and principles involved were not discovered much earlier, but isolated theorems do not make a theory, any more than a pile of bricks makes a building. The materials for our building have been contributed by many different workmen from the days of Euclid down to the present time. Thus, the notion of four harmonic points was familiar to the ancients, who considered it from the metrical point of view as the division of a line internally and externally in the same ratio[1] the involution of six points cut out by any transversal which intersects the sides of a complete quadrilateral as studied by Pappus[2]; but these notions were not made the foundation for any general theory. Taken by themselves, they are of small consequence; it is their relation to other theorems and sets of theorems that gives them their importance. The ancients were doubtless familiar with the theorem, *Two lines determine a point, and two points determine a line*, but they had no glimpse of the wonderful law of duality, of which this theorem is a simple example. The principle of projection, by which many properties of the conic sections may be inferred from corresponding properties of the circle which forms the base of the cone from which they are cut—a principle so natural to modern mathematicians—seems not to have occurred to the Greeks. The ellipse, the hyperbola, and the parabola were to them entirely different curves, to be treated separately with methods appropriate to each. Thus the focus of the ellipse was discovered some five hundred years before the focus of the parabola! It was not till 1522 that Verner[3] of Nürnberg undertook to demonstrate the properties of the conic sections by means of the circle.

162. Unifying principles. In the early years of the seventeenth century—that wonderful epoch in the history of the world which produced a Galileo, a Kepler, a Tycho Brahe, a Descartes, a Desargues, a Pascal, a Cavalieri, a Wallis, a Fermat, a Huygens, a Bacon, a Napier, and a goodly array of lesser lights, to say nothing of a Rembrandt or of a Shakespeare—there began to appear certain unifying principles connecting the great mass of material dug out by the ancients. Thus, in 1604 the great astronomer Kepler[4] introduced the notion that parallel lines should be considered as meeting at an infinite distance, and that a parabola is at once the limiting case of an ellipse and of a hyperbola. He also attributes to the parabola a "blind focus" (*caecus focus*) at infinity on the axis.

163. Desargues. In 1639 Desargues,[5] an architect of Lyons, published a little treatise on the conic sections, in which appears the theorem upon which we have founded the theory of four harmonic points (§ 25). Desargues, however, does not make use of it for that purpose. Four harmonic points are for him a special case of six points in involution when two of the three pairs coincide giving double points. His development of the theory of involution is also different from the purely geometric one which we have adopted, and is based on the theorem (§ 142) that the product of the distances of two conjugate points from the center is constant. He also proves the projective character of an involution of points by showing that when six lines pass through a point and through six points in involution, then any transversal must meet them in six points which are also in involution.

164. Poles and polars. In this little treatise is also contained the theory of poles and polars. The polar line is called a *traversal.*[6] The harmonic properties of poles and polars are given, but Desargues seems not to have arrived at the metrical properties which result when the infinite elements of the plane are introduced. Thus he says, "When the *traversal* is at an infinite distance, all is unimaginable."

165. Desargues's theorem concerning conics through four points. We find in this little book the beautiful theorem concerning a quadrilateral inscribed in a conic section, which is given by his name in § 138. The theorem is not given in terms of a system of conics through four points, for Desargues had no conception of any such system. He states the theorem, in effect, as follows: *Given a simple quadrilateral inscribed in a conic section, every transversal meets the conic and the four sides of the quadrilateral in six points which are in involution.*

166. Extension of the theory of poles and polars to space. As an illustration of his remarkable powers of generalization, we may note that Desargues extended the notion of poles and polars to space of three dimensions for the sphere and for certain other surfaces of the second degree. This is a matter which has not been touched on in this book, but the notion is not difficult to grasp. If we draw through any point *P* in space a line to cut a sphere in two points, *A* and *S*, and then construct the fourth harmonic of *P* with respect to *A* and *B*, the locus of this fourth harmonic, for various lines through *P*, is a plane called the *polar plane* of *P* with respect to the sphere. With this definition and theorem one can easily find dual relations between points and planes in space analogous to those between points and lines in a plane. Desargues closes his discussion of this matter with the remark, "Similar properties may be found for those other solids which are related to the sphere in the same way that the conic section is to the circle." It should not be inferred from this remark, however, that he was acquainted with all the different varieties of surfaces of the second order. The ancients were well acquainted with the surfaces obtained by revolving an ellipse or a parabola about an axis. Even the hyperboloid of two sheets, obtained by revolving the hyperbola about its major axis, was known to them, but probably not the hyperboloid of one sheet, which results from revolving a hyperbola about the other axis. All the other solids of the second degree were probably unknown until their discovery by Euler.[7]

167. Desargues had no conception of the conic section of the locus of intersection of corresponding rays of two projective pencils of rays. He seems to have tried to describe the curve by means of a pair of compasses, moving one leg back and forth along a straight line instead of holding it fixed as in drawing a circle. He does not attempt to define the law of the movement necessary to obtain a conic by this means.

168. Reception of Desargues's work. Strange to say, Desargues's immortal work was heaped with the most violent abuse and held up to ridicule and scorn! "Incredible errors! Enormous mistakes and falsities! Really it is impossible for anyone who is familiar with the science concerning which he wishes to retail his thoughts, to keep from laughing!" Such were the comments of reviewers and critics. Nor were his detractors altogether ignorant and uninstructed men. In spite of the devotion of his pupils and in spite of the admiration and friendship of men like Descartes, Fermat, Mersenne, and Roberval, his book disappeared so completely that two centuries after the date of its publication, when the French geometer Chasles wrote his history of geometry, there was no means of estimating the value of the work done by Desargues. Six years later, however, in 1845, Chasles found a manuscript copy of the "Bruillon-project," made by Desargues's pupil, De la Hire.

169. Conservatism in Desargues's time. It is not necessary to suppose that this effacement of Desargues's work for two centuries was due to the savage attacks of his critics. All this was in accordance with the fashion of the time, and no man escaped bitter denunciation who attempted to improve on the methods of the ancients. Those were days when men refused to believe that a heavy body falls at the same rate as a lighter one, even when Galileo made them see it with their own eyes at the foot of the tower of Pisa. Could they not turn to the exact page and line of Aristotle which declared that the heavier body must fall the faster! "I have read Aristotle's writings from end to end, many times," wrote a Jesuit provincial to the mathematician and astronomer, Christoph Scheiner, at Ingolstadt, whose telescope seemed to reveal certain mysterious spots on the sun, "and I can assure you I have nowhere found anything similar to what you describe. Go, my son, and tranquilize yourself; be assured that what you take for spots on the sun are the faults of your glasses, or of your eyes." The dead hand of Aristotle barred the advance in every department of research. Physicians would have nothing to do with Harvey's discoveries about the circulation of the blood. "Nature is accused of tolerating a vacuum!" exclaimed a priest when Pascal began his experiments on the Puy-de-Dome to show that the column of mercury in a glass tube varied in height with the pressure of the atmosphere.

170. Desargues's style of writing. Nevertheless, authority counted for less at this time in Paris than it did in Italy, and the tragedy enacted in Rome when Galileo was forced to deny his inmost convictions at the bidding of a brutal Inquisition could not have been staged in France. Moreover, in the little company of scientists of which Desargues was a member the utmost liberty of thought and expression was maintained. One very good reason for the disappearance of the work of Desargues is to be found in his style of writing. He failed to heed the very good advice given him in a letter from his warm admirer Descartes.[8] "You may have two designs, both very good and very laudable, but which do not require the same method of procedure: The one is to write for the learned, and show them some new properties of the conic sections which they do not already know; and the other is to write for the curious unlearned, and to do it so that this matter which until now has been understood by only a very few, and which is nevertheless very useful for perspective, for painting, architecture, etc., shall become common and easy to all who wish to study them in your book. If you have the first idea, then it seems to me that it is necessary to avoid using new terms; for the learned are already accustomed to using those of Apollonius, and will not readily change them for others, though better, and thus yours will serve only to render your demonstrations more difficult, and to turn away your readers from your book. If you have the second plan in mind, it is certain that your terms, which are French, and conceived with spirit and grace, will be better received by persons not preoccupied with those of the ancients.... But, if you have that intention, you should make of it a great volume; explain it all so fully and so distinctly that those gentlemen who cannot study without yawning; who cannot distress their imaginations enough to grasp a proposition in geometry, nor turn the leaves of a book to look at the letters in a figure, shall find nothing in your discourse more difficult to understand than the description of an enchanted palace in a fairy story." The point of these remarks is apparent when we note that Desargues introduced some seventy new terms in his little book, of which only one, *involution*, has survived. Curiously enough, this is the one term singled out for the sharpest criticism and ridicule by his reviewer, De Beaugrand.[9] That Descartes knew the character of Desargues's audience better than he did is also evidenced by the fact that De Beaugrand exhausted his patience in reading the first ten pages of the book.

171. Lack of appreciation of Desargues. Desargues's methods, entirely different from the analytic methods just then being developed by Descartes and Fermat, seem to have been little understood. "Between you and me," wrote Descartes[10] to Mersenne, "I can hardly form an idea of what he may have written concerning conics." Desargues seems to have boasted that he owed nothing to any man, and that all his results had come from his own mind. His favorite pupil, De la Hire, did not realize the extraordinary simplicity and generality of his work. It is a remarkable fact that the only one of all his associates to understand and appreciate the methods of Desargues should be a lad of sixteen years!

172. Pascal and his theorem. One does not have to believe all the marvelous stories of Pascal's admiring sisters to credit him with wonderful precocity. We have the fact that in 1640, when he was sixteen years old, he published a little placard, or poster, entitled "Essay pour les conique,"[11] in which his great theorem appears for the first time. His manner of putting it may be a little puzzling to one who has only seen it in the form given in this book, and it may be worth while for the student to compare the two methods of stating it. It is given as follows: *"If in the plane of M, S, Q we draw through M the two lines MK and MV, and through the point S the two lines SK and SV, and let K be the intersection of MK and SK; V the intersection of MV and SV; A the intersection of MA and SA (A is the intersection of SV and MK), and μ the intersection of MV and SK; and if through two of the four points A, K, μ, V, which are not in the same straight line with M and S, such as K and V, we pass the circumference of a circle cutting the lines MV, MP, SV, SK in the points O, P, Q, N; I say that the lines MS, NO, PQ are of the same order."* (By "lines of the same order" Pascal means lines which meet in the same point or are parallel.) By projecting the figure thus described upon another plane he is able to state his theorem for the case where the circle is replaced by any conic section.

173. It must be understood that the "Essay" was only a résumé of a more extended treatise on conics which, owing partly to Pascal's extreme youth, partly to the difficulty of publishing scientific works in those days, and also to his later morbid interest in religious matters, was never published. Leibniz[12] examined a copy of the complete work, and has reported that the great theorem on the mystic hexagram was made the basis of the whole theory, and that Pascal had deduced some four hundred corollaries from it. This would indicate that here was a man able to take the unconnected materials of projective geometry and shape them into some such symmetrical edifice as we have to-day. Unfortunately for science, Pascal's early death prevented the further development of the subject at his hands.

174. In the "Essay" Pascal gives full credit to Desargues, saying of one of the other propositions, "We prove this property also, the original discoverer of which is M. Desargues, of Lyons, one of the greatest minds of this age ... and I wish to acknowledge that I owe to him the little which I have discovered." This acknowledgment led Descartes to believe that Pascal's theorem should also be credited to Desargues. But in the scientific club which the young

Pascal attended in company with his father, who was also a scientist of some reputation, the theorem went by the name of 'la Pascalia,' and Descartes's remarks do not seem to have been taken seriously, which indeed is not to be wondered at, seeing that he was in the habit of giving scant credit to the work of other scientific investigators than himself.

175. De la Hire and his work. De la Hire added little to the development of the subject, but he did put into print much of what Desargues had already worked out, not fully realizing, perhaps, how much was his own and how much he owed to his teacher. Writing in 1679, he says,[13] "I have just read for the first time M. Desargues's little treatise, and have made a copy of it in order to have a more perfect knowledge of it." It was this copy that saved the work of his master from oblivion. De la Hire should be credited, among other things, with the invention of a method by which figures in the plane may be transformed into others of the same order. His method is extremely interesting, and will serve as an exercise for the student in synthetic projective geometry. It is as follows: *Draw two parallel lines, a and b, and select a point P in their plane. Through any point M of the plane draw a line meeting a in A and b in B. Draw a line through B parallel to AP, and let it meet MP in the point M'. It may be shown that the point M' thus obtained does not depend at all on the particular ray MAB used in determining it, so that we have set up a one-to-one correspondence between the points M and M' in the plane.* The student may show that as *M* describes a point-row, *M'* describes a point-row projective to it. As *M* describes a conic, *M'* describes another conic. This sort of correspondence is called a *collineation.* It will be found that the points on the line *b* transform into themselves, as does also the single point *P*. Points on the line *a* transform into points on the line at infinity. The student should remove the metrical features of the construction and take, instead of two parallel lines *a* and *b*, any two lines which may meet in a finite part of the plane. The collineation is a special one in that the general one has an invariant triangle instead of an invariant point and line.

176. Descartes and his influence. The history of synthetic projective geometry has little to do with the work of the great philosopher Descartes, except in an indirect way. The method of algebraic analysis invented by him, and the differential and integral calculus which developed from it, attracted all the interest of the mathematical world for nearly two centuries after

Desargues, and synthetic geometry received scant attention during the rest of the seventeenth century and for the greater part of the eighteenth century. It is difficult for moderns to conceive of the richness and variety of the problems which confronted the first workers in the calculus. To come into the possession of a method which would solve almost automatically problems which had baffled the keenest minds of antiquity; to be able to derive in a few moments results which an Archimedes had toiled long and patiently to reach or a Galileo had determined experimentally; such was the happy experience of mathematicians for a century and a half after Descartes, and it is not to be wondered at that along with this enthusiastic pursuit of new theorems in analysis should come a species of contempt for the methods of the ancients, so that in his preface to his "Méchanique Analytique," published in 1788, Lagrange boasts, "One will find no figures in this work." But at the close of the eighteenth century the field opened up to research by the invention of the calculus began to appear so thoroughly explored that new methods and new objects of investigation began to attract attention. Lagrange himself, in his later years, turned in weariness from analysis and mechanics, and applied himself to chemistry, physics, and philosophical speculations. "This state of mind," says Darboux,[14] "we find almost always at certain moments in the lives of the greatest scholars." At any rate, after lying fallow for almost two centuries, the field of pure geometry was attacked with almost religious enthusiasm.

177. Newton and Maclaurin. But in hastening on to the epoch of Poncelet and Steiner we should not omit to mention the work of Newton and Maclaurin. Although their results were obtained by analysis for the most part, nevertheless they have given us theorems which fall naturally into the domain of synthetic projective geometry. Thus Newton's "organic method"[15] of generating conic sections is closely related to the method which we have made use of in Chapter III. It is as follows: *If two angles, AOS and AO'S, of given magnitudes turn about their respective vertices, O and O', in such a way that the point of intersection, S, of one pair of arms always lies on a straight line, the point of intersection, A, of the other pair of arms will describe a conic.* The proof of this is left to the student.

178. Another method of generating a conic is due to Maclaurin.[16] The construction, which we also leave for the student to justify, is as follows: *If a* *triangle C'PQ move in such a way that its sides, PQ, QC', and C'P, turn around three fixed points, R, A, B, respectively, while two of its vertices, P, Q, slide along two fixed lines, CB' and CA', respectively, then the remaining vertex will describe a conic.*

179. Descriptive geometry and the second revival. The second revival of pure geometry was again to take place at a time of great intellectual activity. The period at the close of the eighteenth and the beginning of the nineteenth century is adorned with a glorious list of mighty names, among which are Gauss, Lagrange, Legendre, Laplace, Monge, Carnot, Poncelet, Cauchy, Fourier, Steiner, Von Staudt, Möbius, Abel, and many others. The renaissance may be said to date from the invention by Monge[17] of the theory of *descriptive geometry*. Descriptive geometry is concerned with the representation of figures in space of three dimensions by means of space of two dimensions. The method commonly used consists in projecting the space figure on two planes (a vertical and a horizontal plane being most convenient), the projections being made most simply for metrical purposes from infinity in directions perpendicular to the two planes of projection. These two planes are then made to coincide by revolving the horizontal into the vertical about their common line. Such is the method of descriptive geometry which in the hands of Monge acquired wonderful generality and elegance. Problems concerning fortifications were worked so quickly by this method that the commandant at the military school at Mézières, where Monge was a draftsman and pupil, viewed the results with distrust. Monge afterward became professor of mathematics at Mézières and gathered around him a group of students destined to have a share in the advancement of pure geometry. Among these were Hachette, Brianchon, Dupin, Chasles, Poncelet, and many others.

180. Duality, homology, continuity, contingent relations. Analytic geometry had left little to do in the way of discovery of new material, and the mathematical world was ready for the construction of the edifice. The activities of the group of men that followed Monge were directed toward this end, and we now begin to hear of the great unifying notions of duality, homology, continuity, contingent relations, and the like. The devotees of pure geometry were beginning to feel the need of a basis for their science which should be at once as general and as rigorous as that of the analysts. Their dream was the building up of a system of geometry which should be independent of analysis. Monge, and after him Poncelet, spent much thought on the so-called "principle of continuity," afterwards discussed by Chasles under the name of the "principle of contingent relations." To get a clear idea of this principle, consider a theorem in geometry in the proof of which certain auxiliary elements are employed. These elements do not appear in the statement of the theorem, and the theorem might possibly be proved without them. In drawing the figure for the proof of the theorem, however, some of these elements may not appear, or, as the analyst would say, they become imaginary. "No matter," says the principle of contingent relations, "the theorem is true, and the proof is valid whether the elements used in the proof are real or imaginary."

181. Poncelet and Cauchy. The efforts of Poncelet to compel the acceptance of this principle independent of analysis resulted in a bitter and perhaps fruitless controversy between him and the great analyst Cauchy. In his review of Poncelet's great work on the projective properties of figures[18] Cauchy says, "In his preliminary discourse the author insists once more on the necessity of admitting into geometry what he calls the 'principle of continuity.' We have already discussed that principle ... and we have found that that principle is, properly speaking, only a strong induction, which cannot be indiscriminately applied to all sorts of questions in geometry, nor even in analysis. The reasons which we have given as the basis of our opinion are not affected by the considerations which the author has developed in his Traité des Propriétés Projectives des Figures." Although this principle is constantly made use of at the present day in all sorts of investigations, careful geometricians are in agreement with Cauchy in this matter, and use it only as a convenient working tool for purposes of exploration. The one-to-one correspondence between geometric forms and algebraic analysis is subject to many and important exceptions. The field of analysis is much more general than the field of geometry, and while there may be a clear notion in analysis to, correspond to every notion in geometry, the opposite is not true. Thus, in analysis we can deal with four coördinates as well as with three, but the existence of a space of four dimensions to correspond to it does not therefore follow. When the geometer speaks of the two real or imaginary intersections of a straight line with a conic, he is really speaking the language of algebra. *Apart from the algebra involved*, it is the height of absurdity to try to distinguish between the two points in which a line *fails to meet a conic!*

182. The work of Poncelet. But Poncelet's right to the title "The Father of Modern Geometry" does not stand or fall with the principle of contingent relations. In spite of the fact that he considered this principle the most important of all his discoveries, his reputation rests on more solid foundations. He was the first to study figures *in homology*, which is, in effect, the collineation described in § 175, where corresponding points lie on straight lines through a fixed point. He was the first to give, by means of the theory of poles and polars, a transformation by which an element is transformed into another of a different sort. Point-to-point transformations will sometimes generalize a theorem, but the transformation discovered by Poncelet may throw a theorem into one of an entirely different aspect. The principle of duality, first stated in definite form by Gergonne,[19] the editor of the mathematical journal in which Poncelet published his researches, was based by Poncelet on his theory of poles and polars. He also put into definite form the notions of the infinitely distant elements in space as all lying on a plane at infinity.

183. The debt which analytic geometry owes to synthetic geometry. The reaction of pure geometry on analytic geometry is clearly seen in the development of the notion of the *class* of a curve, which is the number of tangents that may be drawn from a point in a plane to a given curve lying in that plane. If a point moves along a conic, it is easy to show—and the student is recommended to furnish the proof—that the polar line with respect to a conic remains tangent to another conic. This may be expressed by the statement that the conic is of the second order and also of the second class. It might be thought that if a point moved along a cubic curve, its polar line with respect to a conic would remain tangent to another cubic curve. This is not the case, however, and the investigations of Poncelet and others to determine the class of a given curve were afterward completed by Plücker. The notion of geometrical transformation led also to the very important developments in the theory of invariants, which, geometrically, are the elements and configurations which are not affected by the transformation. The anharmonic ratio of four points is such an invariant, since it remains unaltered under all projective transformations.

184. Steiner and his work. In the work of Poncelet and his contemporaries, Chasles, Brianchon, Hachette, Dupin, Gergonne, and others, the anharmonic ratio enjoyed a fundamental rôle. It is made also the basis of the great work of Steiner,[20] who was the first to treat of the conic, not as the projection of a circle, but as the locus of intersection of corresponding rays of two projective pencils. Steiner not only related to each other, in one-to-one correspondence, point-rows and pencils and all the other fundamental forms, but he set into correspondence even curves and surfaces of higher degrees. This new and fertile conception gave him an easy and direct route into the most abstract and difficult regions of pure geometry. Much of his work was given without any indication of the methods by which he had arrived at it, and many of his results have only recently been verified.

185. Von Staudt and his work. To complete the theory of geometry as we have it to-day it only remained to free it from its dependence on the semimetrical basis of the anharmonic ratio. This work was accomplished by Von Staudt,[21] who applied himself to the restatement of the theory of geometry in a form independent of analytic and metrical notions. The method which has been used in Chapter II to develop the notion of four harmonic points by means of the complete quadrilateral is due to Von Staudt. His work is characterized by a most remarkable generality, in that he is able to discuss real and imaginary forms with equal ease. Thus he assumes a one-to-one correspondence between the points and lines of a plane, and defines a conic as the locus of points which lie on their corresponding lines, and a pencil of rays of the second order as the system of lines which pass through their corresponding points. The point-row and pencil of the second order may be real or imaginary, but his theorems still apply. An illustration of a correspondence of this sort, where the conic is imaginary, is given in § 15 of the first chapter. In defining conjugate imaginary points on a line, Von Staudt made use of an involution of points having no double points. His methods, while elegant and powerful, are hardly adapted to an elementary course, but Reye[22] and others have done much toward simplifying his presentation.

186. Recent developments. It would be only confusing to the student to attempt to trace here the later developments of the science of protective geometry. It is concerned for the most part with curves and surfaces of a higher degree than the second. Purely synthetic methods have been used with marked success in the study of the straight line in space. The struggle between analysis and pure geometry has long since come to an end. Each has its distinct advantages, and the mathematician who cultivates one at the expense of the other will never attain the results that he would attain if both methods were equally ready to his hand. Pure geometry has to its credit some of the finest discoveries in mathematics, and need not apologize for having been born. The day of its usefulness has not passed with the invention of abridged notation and of short methods in analysis. While we may be certain that any geometrical problem may always be stated in analytic form, it does not follow that that statement will be simple or easily interpreted. For many mathematicians the geometric intuitions are weak, and for such the method will have little attraction. On the other hand, there will always be those for whom the subject will have a peculiar glamor—who will follow with delight the curious and unexpected relations between the forms of space. There is a corresponding pleasure, doubtless, for the analyst in tracing the marvelous connections between the various fields in which he wanders, and it is as absurd to shut one's eyes to the beauties in one as it is to ignore those in the other. "Let us cultivate geometry, then," says Darboux,[23] "without wishing in all points to equal it to its rival. Besides, if we were tempted to neglect it, it would not be long in finding in the applications of mathematics, as once it has already done, the means of renewing its life and of developing itself anew. It is like the Giant Antaeus, who renewed, his strength by touching the earth."

Footnotes

1. The more general notion of *anharmonic ratio*, which includes the harmonic ratio as a special case, was also known to the ancients. While we have not found it necessary to make use of the anharmonic ratio in building up our theory, it is so frequently met with in treatises on geometry that some account of it should be given.

Consider any four points, *A*, *B*, *C*, *D*, on a line, and join them to any point *S* not on that line. Then the triangles *ASB*, *GSD*, *ASD*, *CSB*, having all the same altitude, are to each other as their bases. Also, since the area of any triangle is one half the product of any two of its sides by the sine of the angle included between them, we have

$$\frac{AB \times CD}{AD \times CB} = \frac{AS \times BS \sin ASB \times CS \times DS \sin CSD}{AS \times DS \sin ASD \times CS \times BS \sin CSB} = \frac{\sin ASB \times \sin CSD}{\sin ASD \times \sin CSB}$$

Now the fraction on the right would be unchanged if instead of the points *A*, *B*, *C*, *D* we should take any other four points *A'*, *B'*, *C'*, *D'* lying on any other line cutting across *SA*, *SB*, *SC*, *SD*. In other words, *the fraction on the left is unaltered in value if the points A, B, C, D are replaced by any other four points perspective to them.* Again, the fraction on the left is unchanged if some other point were taken instead of *S*. In other words, *the fraction on the right is unaltered if we replace the four lines SA, SB, SC, SD by any other four lines perspective to them.* The fraction on the left is called the *anharmonic ratio* of the four points *A*, *B*, *C*, *D*; the fraction on the right is called the *anharmonic ratio* of the four lines *SA*, *SB*, *SC*, *SD*. The anharmonic ratio of four points is sometimes written *(ABCD)*, so that

$$\frac{AB \times CD}{AD \times CB} = (ABCD).$$

If we take the points in different order, the value of the anharmonic ratio will not necessarily remain the same. The twenty-four different ways of writing them will, however, give not more than six different values for the anharmonic ratio, for by writing out the fractions which define them we can find that *(ABCD)* = *(BADC)* = *(CDAB)* = *(DCBA)*. If we write *(ABCD)* = *a*, it is not difficult to show that the six values are

$$a; 1/a; 1-a; 1/(1-a); (a-1)/a; a/(a-1).$$

The proof of this we leave to the student.

If *A*, *B*, *C*, *D* are four harmonic points (see Fig. 6, p. *22), and a quadrilateral *KLMN* is constructed such that *KL* and *MN* pass through *A*, *KN* and *LM* through *C*, *LN* through *B*, and *KM* through *D*, then, projecting *A*, *B*, *C*, *D* from *L* upon *KM*, we have *(ABCD)* = *(KOMD)*, where *O* is the intersection of *KM* with *LN*. But, projecting again the points *K*, *O*, *M*, *D* from *N* back upon the line *AB*, we have *(KOMD)* = *(CBAD)*. From this we have

$$(ABCD) = (CBAD),$$

or

$$a = a/(a - 1);$$

whence $a = 0$ or $a = 2$. But it is easy to see that $a = 0$ implies that two of the four points coincide. For four harmonic points, therefore, the six values of the anharmonic ratio reduce to three, namely, 2, $\frac{1}{2}$, and -1. Incidentally we see that if an interchange of any two points in an anharmonic ratio does not change its value, then the four points are harmonic.

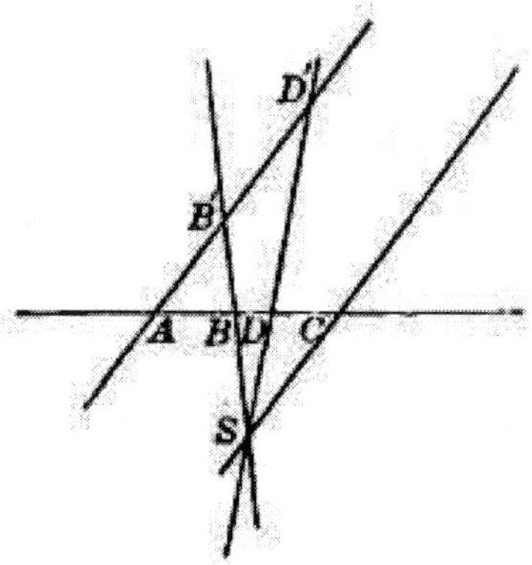

FIG. 49

Many theorems of projective geometry are succinctly stated in terms of anharmonic ratios. Thus, the *anharmonic ratio of any four elements of a form is equal to the anharmonic ratio of the corresponding four elements in any form projectively related to it. The anharmonic ratio of the lines joining any four fixed points on a conic to a variable fifthpoint on the conic is constant. The locus of points from which four points in a plane are seen along four rays of constant anharmonic ratio is a conic through the four points.* We leave these theorems for the student, who may also justify the following solution of the problem: *Given three points and a certain anharmonic ratio, to find a fourth point which shall have with the given three the given anharmonic ratio.* Let *A*, *B*, *D* be the three given points (Fig. 49). On any convenient line through *A* take two points B' and D' such that AB'/AD' is equal to the given anharmonic ratio. Join BB' and DD' and let the two lines meet in *S*. Draw through *S* a parallel to AB'. This line will meet *AB* in the required point *C*.

2. Pappus, Mathematicae Collectiones, vii, 129.

3. J. Verneri, Libellus super vigintiduobus elementis conicis, etc. 1522.

4. Kepler, Ad Vitellionem paralipomena quibus astronomiae pars optica traditur. 1604.

5. Desargues, Bruillon-project d'une atteinte aux événements des rencontres d'un cône avec un plan. 1639. Edited and analyzed by Poudra, 1864.

6. The term 'pole' was first introduced, in the sense in which we have used it, in 1810, by a French mathematician named Servois (Gergonne, *Annales des Mathéématiques*, I, 337), and the corresponding term 'polar' by the editor, Gergonne, of this same journal three years later.

7. Euler, Introductio in analysin infinitorum, Appendix, cap. V. 1748.

8. Œuvres de Desargues, t. II, 132.

9. Œuvres de Desargues, t. II, 370.

10. Œuvres de Descartes, t. II, 499.

11. Œuvres de Pascal, par Brunsehvig et Boutroux, t. I, 252.

12. Chasles, Histoire de la Géométrie, 70.

13. Œuvres de Desargues, t. I, 231.

14. See Ball, History of Mathematics, French edition, t. II, 233.

15. Newton, Principia, lib. i, lemma XXI.

16. Maclaurin, Philosophical Transactions of the Royal Society of London, 1735.

17. Monge, Géométrie Descriptive. 1800.

18. Poncelet, Traité des Propriétés Projectives des Figures. 1822. (See p. 357, Vol. II, of the edition of 1866.)

19. Gergonne, *Annales de Mathématiques, XVI, 209. 1826.*

20. Steiner, Systematische Ehtwickelung der Abhängigkeit geometrischer Gestalten von einander. 1832.

21. Von Staudt, Geometrie der Lage. 1847.

22. Reye, Geometrie der Lage. Translated by Holgate, 1897.

23. Ball, loc. cit. p. 261.

Made in the USA
Monee, IL
08 November 2026

67377452R00065